BASIC ELEMENTS OF
CRYSTALLOGRAPHY

BASIC ELEMENTS OF CRYSTALLOGRAPHY

2nd Edition

Nevill Gonzalez Szwacki
Teresa Szwacka

PAN STANFORD PUBLISHING

Published by

Pan Stanford Publishing Pte. Ltd.
Penthouse Level, Suntec Tower 3
8 Temasek Boulevard
Singapore 038988

Email: editorial@panstanford.com
Web: www.panstanford.com

British Library Cataloguing-in-Publication Data
A catalogue record for this book is available from the British Library.

Basic Elements of Crystallography (2nd Edition)

ISBN 978-981-4613-57-6 (Hardcover)
ISBN 978-981-4613-58-3 (eBook)

Printed in the USA

Contents

Preface

It has been five years since the first edition of this book, which offers the readers a friendly look inside crystallography. In the present edition, we have added a new chapter and updated some of the existing ones. Especially the first chapter now offers all the basic concepts of crystallography presented on the example of simple structures. Most importantly, perhaps, we have provided the book with solutions to almost all exercises.

This new edition (as the old one) is intended to be a complete and clear introduction to the field of crystallography for undergraduate and graduate students and lecturers in physics, chemistry, biology, materials and earth sciences, or engineering. It includes an extensive discussion of the 14 Bravais lattices and the reciprocal to them, basic concepts of point- and space-group symmetries, the crystal structure of elements and binary compounds, and much more. Besides that, the reader can find up-to-date values for the lattice constants of most elements and about 650 binary compounds (half of them containing rare earth metals). The entire notation in this book is consistent with the International Tables for Crystallography.

We have improved the quality of many illustrations. Our purpose remains to show rather than describe "using many words" the structure of materials and its basic properties. We believe that even readers who are completely not familiar with the topic, but still want to learn how the atoms are arranged in crystal structures, will find this book useful.

The text is organized into seven chapters. Chapter 1 introduces basic concepts and definitions in the field of crystallography, starting with one- and two-dimensional structures. Chapter 2 provides a detailed description of the 14 Bravais lattices. Chapter 3 describes the most important crystal structures of the elements with special emphasis on the close-packed structures and the interstices present in them. Chapter 4 presents the structures of important binary compounds and reports the lattice constants of about 650 of them. Chapter 5 is devoted to the reciprocal lattice. The relation between the direct lattice and its reciprocal is explained in Chapter 6. Finally, the basic concepts of X-ray diffraction are introduced in Chapter 7, which is the new chapter added to the present edition. All the chapters are accompanied by exercises designed in a way that encourages students to explore the crystal structures they are learning about. Detailed solutions to most exercises can be found in Appendix A.

We invite the reader to make use of crystallographic databases. In most of the database web pages, it is possible to visualize crystal structures in 3D either directly from the web browser or by downloading input files with the coordinates of the structures and using, e.g., the open-source software called Jmol (http://www.jmol.org). Some of the freely available (or with open access options) databases are

- Inorganic Crystal Structure Database (ICSD) (http://icsd.ill.fr)
- American Mineralogist Crystal Structure Database (AMCSD) (http://rruff.geo.arizona.edu/AMS/amcsd.php)
- Crystallography Open Database (COD) (http://cod.ibt.lt).

Last but not least, we thank all the reviewers of the first edition for their valuable inputs and hope that this revised edition will be welcomed by the readers and continue to be a gateway for understanding more advanced texts on this topic.

Nevill Gonzalez Szwacki
Warsaw, Poland

Teresa Szwacka
Mérida, Venezuela

Abbreviations

The following abbreviations are used throughout this book:

bcc	body centered cubic
ccp	cubic close-packed
dhcp	double hexagonal close-packed
fcc	face centered cubic
hcp	hexagonal close-packed
sc	simple cubic
thcp	triple hexagonal close-packed
NN	nearest neighbor
NNN	next nearest neighbor
TNN	third-nearest neighbor
RE	rare earth
TM	transition metal

Chapter 1

One- and Two-Dimensional Crystal Lattices

1.1 Introduction

Many of the materials surrounding us (metals, semiconductors, or insulators) have a crystalline structure. That is to say, they represent a set of atoms distributed in space in a particular way. Strictly speaking, this is the case when the atoms occupy their equilibrium positions. Obviously, in the real case they are vibrating. Below we will see examples of crystal structures, beginning with one-dimensional cases.

1.2 One-Dimensional Crystal Structures

A one-dimensional crystal structure is formed by a set of atoms or groups of them distributed periodically in one direction. In Fig. 1.1 there are three examples of one-dimensional crystal structures. In all three cases, the whole crystal structure may be obtained by placing atoms (or groups of them), at a distance $a = |\vec{a}|$ from one another, along a straight line. When we translate an infinite structure by vector $\vec{a}$ we obtain the same structure. The same will occur if we

Basic Elements of Crystallography (2nd Edition)
Nevill Gonzalez Szwacki and Teresa Szwacka
Copyright © 2016 Pan Stanford Publishing Pte. Ltd.
ISBN 978-981-4613-57-6 (Hardcover), 978-981-4613-58-3 (eBook)
www.panstanford.com

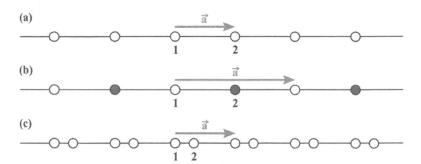

Figure 1.1 Three different one-dimensional crystal structures: (a) periodic repetition of identical atoms, (b) periodic repetition of a building block composed of two different atoms, and (c) periodic repetition of a building block composed of two identical atoms.

translate the structure by a vector equal to the multiple of vector $\vec{a}$, that is, $n\vec{a}$, where $n \in \mathbb{Z}$. The vector $\vec{a}$ is called a primitive translation vector. A clear difference can be seen between the crystal structure from Fig. 1.1a and the other two structures in this figure. In the structure from Fig. 1.1a, all the atoms have equivalent positions in space, while in the case of structures from Figs. 1.1b and 1.1c this does not occur. It can be easily observed that in the structure from Fig. 1.1b the nearest neighbor (NN) atoms of the atom labeled 1 (open circles) are of another type (filed circles) and the NNs of the atom labeled 2 are atoms of type 1. In the case of the structure from Fig. 1.1c, the atom labeled 1 has its NN on the right side, while the atom labeled 2, on the left side.

The fact that after translating an infinite crystal structure by the primitive translation vector $\vec{a}$ or its multiple, $n\vec{a}$, we obtain the same structure characterize all crystal structures. This is the starting point to introduce a certain mathematical abstraction called *lattice*—a periodic arrangement of points in space, whose positions are given by vectors $n\vec{a}$ which can have as an initial point any point of the one-dimensional space. The atomic arrangement in the crystal structure looks the same from any point (node) of the lattice, as can be seen in Fig. 1.2, where we show two different arrangements of lattice points with respect to atoms of the crystal structures from Fig. 1.1. Therefore, all lattice points have equivalent positions in the crystal structure, which we cannot say in general about the

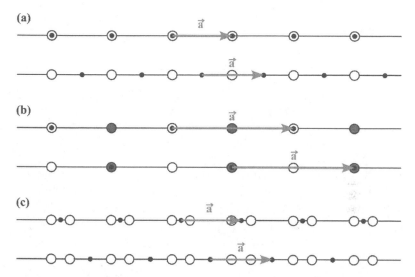

Figure 1.2 Two different arrangements of lattice points with respect to atoms of each of the crystal structures shown in Fig. 1.1. In all cases, the atomic arrangement looks the same from any point of the lattice. The lattice basis vector $\vec{a}$ defines its primitive cell.

atoms. As it is shown, e.g., in Fig. 1.1b, the equivalency between the neighborhoods of the atoms does not exist when the crystal structure is made up of more than one type of atoms. Figure 1.1c shows that the distribution of atoms in space can be another possible source of inequivalency between the atoms. The lattice is a mathematical object that possesses the information about the translation symmetry of the crystal structure. The relation between the structure and its lattice will be discussed in detail below.

Let us now determine the number of atoms in a volume defined by vector $\vec{a}$. When the initial and final points of vector $\vec{a}$ coincide with the center of atoms (see Fig. 1.2a), one-half of each atom belongs to the volume in consideration, so the volume possesses one atom. Besides that, segment a may have other atoms, as shown in Fig. 1.2b,c. The volume defined by vector $\vec{a}$ always contains the same number of atoms, independently of the position of the initial point of the vector.

The primitive translation vector $\vec{a}$, also called the *basis vector* of the lattice, defines a *unit cell* of this lattice, which contains exactly

Figure 1.3 Crystal structure from Fig. 1.1a with two choices of lattice points (different from those shown in Fig. 1.2a) with respect to atoms of the structure. In this case, the basis is composed of two atoms.

one lattice point. This cell is called a *primitive cell* and its "volume" is equal to $a = |\vec{a}|$. From now on, the volume of the primitive unit cell will be denoted by Ω_0. The entire lattice space with all lattice points can be obtained by duplicating the primitive cell an infinite number of times. The position of each cell replica is given by vector $n\vec{a}$. The crystal structure is obtained when we attach to each lattice point a group of atoms, which are within the volume of the primitive cell. This group is called the *basis*. In the case of the crystal structure from Fig. 1.2a the basis consists of one atom, while in the case of Fig. 1.2b,c of two atoms.

It is obvious that there is more than one way to propose a lattice for a certain crystal structure. For example, the lattice shown in Fig. 1.3 could be another option for the structure from Fig. 1.1a. The basis vector of this lattice is two times longer than that defining the lattice proposed in Fig. 1.2a. We can see in Fig. 1.3 that the atomic basis of the structure has now two atoms instead of one we had in the previous case. Of course, in general, we choose a lattice in which the atomic basis of the crystal structure is the smallest one, but sometimes it is convenient to use a different lattice, as we will see further on.

Until now, we have spoken about the translation symmetry of a one-dimensional crystal structure. The three crystal structures shown in Fig. 1.1 possess also *point symmetry*. A point operation is a geometric transformation that leaves at least one point invariant (rotations, reflections, etc.). In the case of a one-dimensional structure, point symmetry operations are limited to identity (whose printed symbol is 1) and a reflection through a point which represents at the same time an inversion through this point. Figure 1.4 shows reflection points (mirror points) or inversion

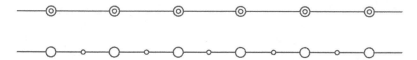

Figure 1.4 Crystal structure from Fig. 1.1a with graphical symbols (small circles) that represent reflection points. The upper drawing shows the reflection points that overlap with the centers of atoms, and the lower drawing shows the set of reflection points that are equidistant from the centers of atoms.

centers of the crystal structure from Fig. 1.1a, using a common graphical symbol for such symmetry elements; their common printed symbol is $\bar{1}$. In the case of crystal structures shown in Figs. 1.2 and 1.3 the lattice points were placed in reflection points of the structure, and the origin of the unit cell coincides, in each case, with a reflection point. Such origin is called a *conventional origin*. If a one-dimensional structure does not possess reflection points, the origin is arbitrary.

1.3 Two-Dimensional Crystal Structures

We will now look at the two-dimensional case, beginning with the example of a two-dimensional crystal structure shown in Fig. 1.5. This structure possesses twofold rotation points represented in the figure by graphical symbols. We can observe in the figure the different positions of the twofold rotation points with respect to the atoms of the structure. The translation symmetry of the structure is illustrated in Fig. 1.6. In this figure, the vectors $\vec{a}$, $\vec{b}$ are primitive translation vectors. If the infinite crystal structure is translated to a vector $\vec{R}$, which is a linear combination of vectors $\vec{a}$ and $\vec{b}$

$$\vec{R} = n_1\vec{a} + n_2\vec{b}, \text{ where } n_1, n_2 \in \mathbb{Z}, \qquad (1.1)$$

then the same structure as the original one is obtained. Vectors $\vec{a}$, $\vec{b}$ can be used to define a lattice. The lattice points may overlap with the centers of atoms like in Figs. 1.5 and 1.6. By translating the replicas of cell I, defined by vectors $\vec{a}$, $\vec{b}$ in Fig. 1.6, through all the vectors $\vec{R}$, we can reproduce the entire lattice space. Cell I, however, is not the only one that can reproduce all the lattice space. There

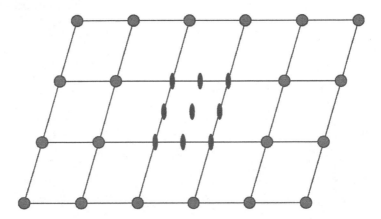

Figure 1.5 Two-dimensional crystal structure. The lattice points overlap with the centers of atoms. In addition, we show different positions of twofold rotation points using graphical symbols of such symmetry elements.

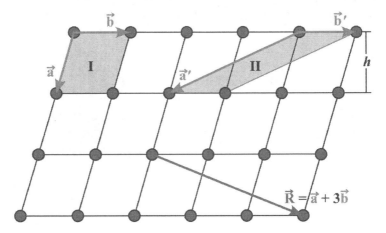

Figure 1.6 Two-dimensional crystal structure. The lattice points overlap with the centers of atoms. Cells I and II are examples of two unit cells that can reproduce the lattice.

is an infinite number of such cells. For example, the cell labeled II in Fig. 1.6, defined by vectors $\vec{a}'$, $\vec{b}'$, can also reproduce the entire lattice. The volumes of cells I and II are

$$\Omega_0 = ba\sin\sphericalangle\left(\vec{a}, \vec{b}\right) = bh \text{ and } \Omega'_0 = b'a'\sin\sphericalangle\left(\vec{a}', \vec{b}'\right) = bh,$$

(1.2)

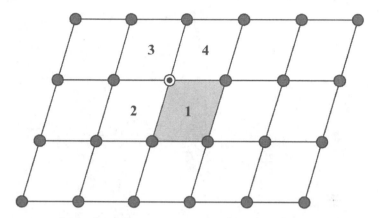

Figure 1.7 Two-dimensional crystal structure. The highlighted atom (lattice point) belongs to four cells which are marked from 1 to 4, therefore only a fraction of this atom (lattice point) belongs to the highlighted cell 1.

respectively, where $b' = b$ and $a \sin \sphericalangle \left(\vec{a}, \vec{b} \right) = a' \sin \sphericalangle \left(\vec{a}', \vec{b}' \right) = h$ [to determine h we have used the trigonometric identity $\sin (\pi - \alpha) = \sin \alpha$]. So, the two volumes are identical.

Let us now demonstrate that cells I and II from Fig. 1.6 are primitive, since they contain only one lattice point. Those cells and also the cell 1 in Fig. 1.7 have 4 atoms at the vertices whose centers represent points of the lattice. Both, the atoms and the lattice points, are shared with neighboring cells. This is shown in Fig. 1.7, where a highlighted atom (lattice point) is shared by cells 1 to 4. Each cell has a fraction of an atom (lattice point) and the sum of the fractions is 1, giving one atom (lattice point) per cell. The points from the vertices of any cell that is a parallelogram contribute exactly with one lattice point to the cell. All primitive cells have the same volume. This volume corresponds to one point of a lattice. The most commonly used primitive cell is the one which is defined by the shortest or one of the shortest primitive translation vectors of the lattice (e.g., $\vec{a}$, $\vec{b}$ from Fig. 1.6). These vectors are called *basis vectors*. Note that the choice of basis vectors is not unique, since even the shortest vectors can be chosen in several different ways. For the discussed two-dimensional lattice, we can also choose a non-primitive unit cell. An example of such a cell is shown in Fig. 1.8. The cell in this

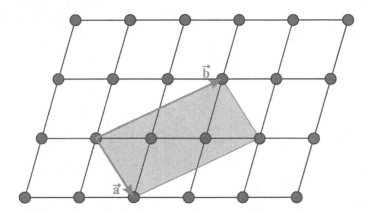

Figure 1.8 Two-dimensional crystal structure. The lattice points overlap with the centers of atoms. The cell shown in the figure is not primitive since contains 3 lattice points.

figure possesses two lattice points inside, so the total number of points belonging to it is three.

We will refer to the unit cell as *conventional cell* if it is chosen such that the symmetry of the crystal structure is displayed best. In the case of the crystal structure shown in Fig. 1.6 this is achieved with a primitive unit cell, but frequently the so-called *centered unit cell* (with more than one lattice point per cell) is necessary to fulfill this condition. Such a centered unit cell should have the same point symmetry as the lattice. The basis vectors that define the conventional cell have to be the shortest possible and the angle between them has to be as close to 90° as possible but greater than or equal to 90°. Additionally, the origin of a conventional cell has to coincide with a center of symmetry or a point of high symmetry of the structure. Cell I in Fig. 1.6 is an example of a conventional cell and the set of twofold rotation points shown in Fig. 1.5 constitutes the diagram of symmetry elements of this structure.

Let us now place an additional atom in the middle of each parallelogram of type I from Fig. 1.6. The resulting structure is shown in Fig. 1.9. The additional atoms are of the same type as the atoms of the original structure. In Fig. 1.9, we can observe that the resulting crystal structure is of the same type as the original one, since the symmetry of both structures is the same (compare the

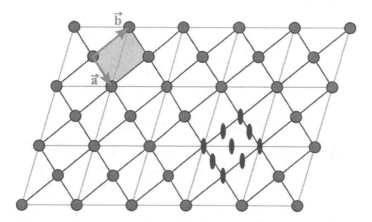

Figure 1.9 A two-dimensional crystal structure obtained from the structure from Fig. 1.6 by placing an additional atom at the center of each unit cell of type I. The lattice points of the resulting structure overlap with the centers of atoms. Vectors $\vec{a}$, $\vec{b}$ define a primitive unit cell which contains one atom. The figure shows also the diagram of symmetry elements (twofold rotation points) of the structure.

diagrams of symmetry elements shown in Figs. 1.5 and 1.9). Vectors $\vec{a}$, $\vec{b}$ in Fig. 1.9 are the primitive translation vectors of a lattice whose nodes overlap with the centers of atoms. They define a conventional cell (note that its origin coincides with a twofold rotation point).

If we place atoms in the middle of each parallelogram of Fig. 1.6 that are of a different type than the atoms of the host structure, then the resulting crystal structure will look as shown in Fig. 1.10. In this case, the smallest atomic basis contains two atoms (one of each type) and the cell of type I from Fig. 1.6 represents the conventional unit cell. It is interesting to observe that the distribution of twofold rotation points in the structures from Figs. 1.6 and 1.10 is the same.

Next, we will consider the case in which we place an additional atom of the same type as the host atoms in the cell of type I from Fig. 1.6; however, this time not in the middle of each parallelogram, but in a position with less symmetry, as shown in Fig. 1.11. In this case, the smallest atomic basis also contains two atoms (like in the previous case) but this time they are of the same type. We can observe in Fig. 1.11 that this crystal structure can be considered as a superposition of two identical crystal substructures, which are

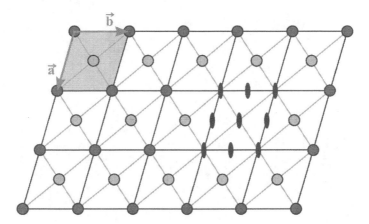

Figure 1.10 A two-dimensional crystal structure made up of two types of atoms. The unit cell, defined by vectors $\vec{a}, \vec{b}$, has 2 atoms. The diagram of symmetry elements (twofold rotation points) of the structure is supplied.

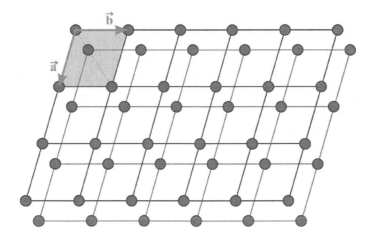

Figure 1.11 A two-dimensional crystal structure that is a superposition of two identical substructures (structures from Fig. 1.6). The lattice points overlap with the centers of atoms of one of the substructures. The primitive cell of the lattice is defined by vectors $\vec{a}, \vec{b}$ and has 2 atoms. Each of the basis atoms belongs to a different substructure.

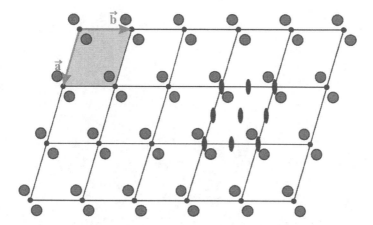

Figure 1.12   Two-dimensional crystal structure. The lattice points overlap with twofold rotation points of the structure, so the unit cell defined by vectors $\vec{a}, \vec{b}$ is conventional. The diagram of symmetry elements of the structure is provided.

structures from Fig. 1.6. Moreover, in Fig. 1.12, we can see that the structure from Fig. 1.11 has the same symmetry as the structure from Fig. 1.6 (compare their diagrams of symmetry elements shown in Figs. 1.12 and 1.5, respectively). The lattice points in Fig. 1.12 coincide with twofold rotation points of the structure. Therefore the unit cell shown there is conventional, while the cell from Fig. 1.11 is not conventional as its origin does not coincide with any of the twofold rotation points of the structure.

We may conclude that all the two-dimensional structures considered by us until now have the same symmetry which coincides, in each case, with the symmetry of the lattice. Their diagrams of symmetry elements are composed only of twofold rotation points. The conventional unit cell is primitive and has its origin at a twofold rotation point.

Finally, Fig. 1.13 shows a structure obtained in a similar way as that from Fig. 1.11, but the additional atom is now of different type than the host atom in the cell of type I from Fig. 1.6. In this way, we have obtained a two-dimensional structure with the lowest possible symmetry. Only onefold rotation points are possible. Since this structure does not possess points of high symmetry (for all

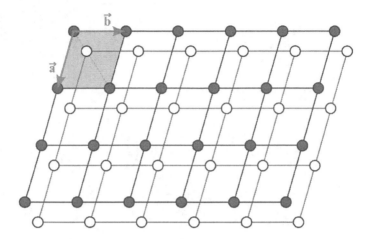

Figure 1.13   A two-dimensional crystal structure made up of two types of atoms. The structure is a superposition of two substructures. Each of them is made up of one type of atoms. The lattice points overlap with the centers of atoms of one of the substructures. The unit cell defined by vectors $\vec{a}, \vec{b}$ is conventional.

points, the symmetry is the same), the unit cell shown in Fig. 1.13 is conventional.

In the following, we will consider two more examples of two-dimensional crystal structures, namely, the honeycomb and the two-dimensional hexagonal structures. Figure 1.14 shows the honeycomb structure with a primitive unit cell that contains two atoms. This is the smallest atomic basis for the honeycomb structure. We considered in Fig. 1.14 two choices for the origin of the unit cell. In each case, the location of the lattice points with respect to the atoms is different (the same was already observed in Figs. 1.11 and 1.12). In one case, the lattice points overlap with the centers of atoms and in the other case, they overlap with the centers of the hexagons. The conventional cell has its origin at a sixfold rotation point of the structure (the center of a hexagon).

In Fig. 1.15, we show a lattice for the honeycomb structure from Fig. 1.14. This lattice is a two-dimensional hexagonal lattice. Vectors $\vec{a}$, $\vec{b}$ defined in Fig. 1.14 are the basis vectors of this lattice and they define a unit cell which has the shape of a rhomb. The points of the lattice shown in Fig. 1.15 are sixfold

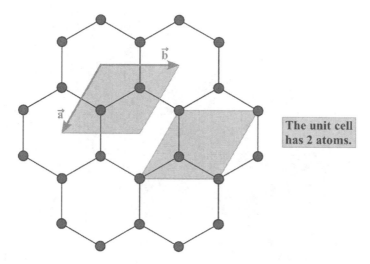

Figure 1.14 The honeycomb structure. In the figure are shown two positions of the origin of the unit cell with respect to the atoms of the structure.

rotation points and the geometric centers of the equilateral triangles (building blocks of the hexagons) are threefold rotation points. If the lattice points of the structure that is shown in Fig. 1.14 overlap with the centers of the honeycombs, then the sixfold rotation points of the lattice overlap with sixfold rotation points of the honeycomb structure. Instead, if we place the lattice points at the centers of atoms, then the sixfold rotation points of the lattice overlap with half of the threefold rotation points of the honeycomb structure and half of the threefold rotation points of the lattice overlap with sixfold rotation points of the honeycomb structure. The symmetry of the honeycomb structure is, of course, displayed best when the lattice points coincide with the centers of the hexagons (honeycombs).

If we now place an additional atom (of the same type) at the center of each hexagon from Fig. 1.14, then the honeycomb structure transforms into a hexagonal structure. In the new structure, the primitive translation vectors can be chosen in the way shown in Fig. 1.16 and the unit cell defined by them is conventional. The atomic basis consists of one atom.

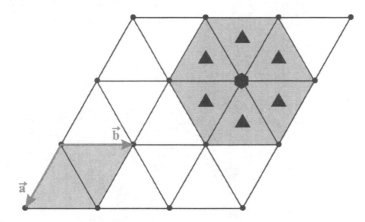

Figure 1.15 A hexagonal lattice for the honeycomb structure shown in Fig. 1.14. The basis vectors $\vec{a}$, $\vec{b}$ are defined in Fig. 1.14. The lattice points are sixfold rotation points and the geometric centers of the equilateral triangles overlap with threefold rotation points of the lattice. In the figure, we also show the graphical symbols for the threefold and sixfold rotation points.

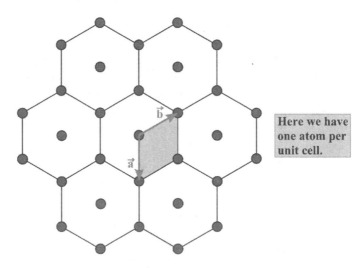

Here we have one atom per unit cell.

Figure 1.16 A two-dimensional hexagonal structure. The primitive unit cell defined by vectors $\vec{a}$, $\vec{b}$ is a conventional cell.

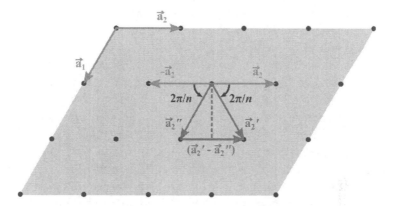

Figure 1.17 A construction made using the basis vector $\vec{a}_2$ and the opposite to it, $-\vec{a}_2$, to show that there are only one-, two-, three-, four-, and sixfold rotation points in a two-dimensional crystal structure or lattice.

Let us now show that the presence of translation symmetry implies that there are only one-, two-, three-, four-, and sixfold rotation points in a two-dimensional crystal structure or lattice. We will explain this using Fig. 1.17. In this figure, we make rotations of the basis vector $\vec{a}_2$ and the opposite to it, $-\vec{a}_2$, by the same angle $2\pi/n$ $(n \in \mathbb{Z})$ but in opposite directions and the difference of the rotated vectors is shown in the figure. The translation symmetry requires that the difference $\left(\vec{a}_2' - \vec{a}_2''\right)$ be a multiple of vector $\vec{a}_2$, which imposes certain condition on the integer number n. We have

$$\begin{cases} \vec{a}_2' - \vec{a}_2'' = m\vec{a}_2 \\ 2a_2 \cos\left(2\pi/n\right) = ma_2 \end{cases}, \text{ where } m \in \mathbb{Z}. \qquad (1.3)$$

From the above we obtain

$$\cos\left(2\pi/n\right) = \frac{1}{2}m \qquad (1.4)$$

and the possible values of integer m and $\cos\left(2\pi/n\right)$ are

$$m = 0, \pm 1, \pm 2 \text{ and } \cos\left(2\pi/n\right) = 0, \pm\frac{1}{2}, \pm 1, \qquad (1.5)$$

respectively. Therefore, from Eq. 1.5 we obtained that the only rotations that can be performed are those by angles

$$\frac{2\pi}{n} = \frac{2\pi}{1}, \frac{2\pi}{2}, \frac{2\pi}{3}, \frac{2\pi}{4}, \frac{2\pi}{6}. \qquad (1.6)$$

From the above, we can finally conclude that in a two-dimensional lattice, there are only allowed one-, two-, three-, four-, and sixfold rotation points.

We will now identify the possible two-dimensional lattices taking into account the limitations for the rotation points described above. The set of all point symmetry operations which carry the lattice (crystal structure) into itself is known as the *point group* of this lattice (crystal structure). We know already that a point operation is a geometric transformation that leaves at least one point invariant. A rotation around a point and a reflection though a line are such operations in two dimensions. We can see, on the basis of the lattices considered here, that it is possible to identify finite volumes of the lattice space which have the same point symmetry as the lattice. Let us think about the smallest such volumes. In the case of the lattices for crystal structures from Figs. 1.6, 1.9, 1.10, and 1.12, these volumes are the primitive cells, which are parallelograms, defined by vectors $\vec{a}$, $\vec{b}$. In the case of the hexagonal lattice, the smallest such volume is the hexagon shown in Fig. 1.15. For the rest of lattice types, the volumes may have only a shape of a rectangle or a square. In Fig. 1.18a,b, we show the twofold rotation points that are at the centers of a parallelogram and a rectangle, respectively. The rotation of 180° around the center of a parallelogram or a rectangle carries each of them into itself. The geometric centers of a square and of a regular hexagon represent fourfold and sixfold rotation points, respectively, that are labeled with the corresponding graphical symbols in Fig. 1.18c,d. With the exception of a parallelogram, the plane figures shown in Fig. 1.18 have, additionally, reflection lines within their sets of symmetry elements.

To identify all possible two-dimensional lattices, we will start with the plane figures shown in Fig. 1.18. Let us place lattice points at the vertices of each plane figure shown in Fig. 1.18a–c and in the case of Fig. 1.18d also at the geometric center of the hexagon. The set of lattice points for each case has the same point symmetry as the corresponding plane figure. In this way, we have obtained four plane lattice types shown in Fig. 1.19a,b,d,e. By placing an additional lattice point in the geometric center of a parallelogram, a rectangle, or a square, we obtain a set of lattice points that also have the point symmetry of the corresponding plane figure. However,

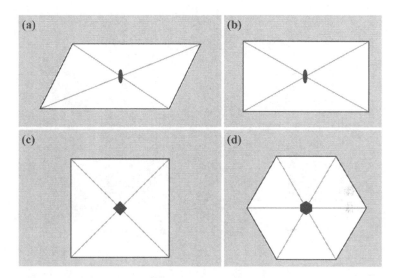

Figure 1.18 Rotation points that overlap with the geometric centers of the following plane figures: (a) parallelogram, (b) rectangle, (c) square, and (d) regular hexagon. The graphical symbols of the rotation points are shown.

only in the case of a rectangle, a new type of lattice is obtained. This lattice, which is shown in Fig. 1.19c, is known as the centered rectangular lattice. In such a lattice, there is no primitive unit cell with the point symmetry of a rectangle, while in the case of a lattice obtained locating the points in the vertices and at the center of a parallelogram (square), we can find a primitive unit cell of the same point symmetry as the symmetry of the unit cell shown in Fig. 1.19a (Fig. 1.19d), which is a parallelogram (square) with half volume of the centered one. In the case of a hexagonal lattice, the placement of additional lattice points that do not change the point symmetry of the lattice does not conduce to any new type of lattice. This case is considered in Exercise 1.4. The presence of the centering point in the rectangular unit cell does not change its point symmetry, but the plane symmetry of the centered rectangular lattice is different from that of the rectangular lattice (see Exercises 1.14 and 1.15).

To conclude, there are five different two-dimensional types of lattices, which are classified in four *crystal systems*: oblique, rectangular, square, and hexagonal. The crystal systems are identified by the point groups of the geometric figures shown in Fig. 1.18,

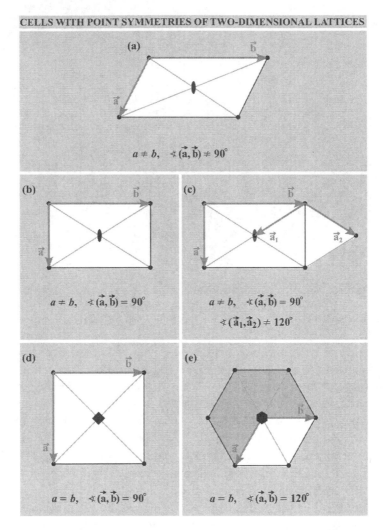

CELLS WITH POINT SYMMETRIES OF TWO-DIMENSIONAL LATTICES

(a)

$a \neq b, \quad \sphericalangle(\vec{a}, \vec{b}) \neq 90°$

(b)

$a \neq b, \quad \sphericalangle(\vec{a}, \vec{b}) = 90°$

(c)

$a \neq b, \quad \sphericalangle(\vec{a}, \vec{b}) = 90°$

$\sphericalangle(\vec{a}_1, \vec{a}_2) \neq 120°$

(d)

$a = b, \quad \sphericalangle(\vec{a}, \vec{b}) = 90°$

(e)

$a = b, \quad \sphericalangle(\vec{a}, \vec{b}) = 120°$

Figure 1.19 Conventional unit cells for the five lattices existing in two dimensions: (a) oblique, (b) rectangular, (c) centered rectangular, (d) square, and (e) hexagonal. In cases (a–d) the *n*-fold rotation point shown in the figure corresponds to both the unit cell and the lattice, while in case (e) it corresponds only to the lattice.

which are the largest point groups belonging to the crystal systems. A parallelogram, rectangle, square, and one-third of a hexagon represent the unit cells of the five lattices. To the rectangular crystal system belongs two lattice types, rectangular and centered rectangular, while to each of the remaining crystal systems belongs only one type of lattice. All of them are shown in Fig. 1.19.

1.4 Crystallographic Point and Space Groups in Two Dimensions

The crystallographic *space group* is formed by the point-group operations (defined in previous section) and translation symmetry operations, which carry the crystal structure into itself. Owing to the presence of an atomic basis in the crystal structure, its point symmetry is in general lower than the point symmetry of its lattice. The point group of the structure represents a subgroup of the point group of its lattice. Before continuing with the description of the crystallographic point and space groups in two dimensions, we will introduce a few new concepts.

The *lattice direction* can be defined as a line joining any two points of the lattice. Vector $\vec{a}_{\bar{1}}$ defined in Fig. 1.20, $\vec{a}_{\bar{1}} = l_1\vec{a} + l_2\vec{b} = 1\cdot\vec{a} + 1\cdot\vec{b}$, is the shortest translation vector along the lattice direction shown in this figure. To denote a lattice direction, we use the nota-

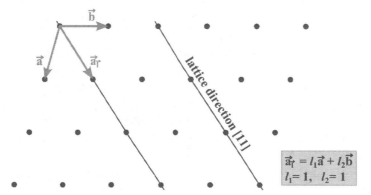

Figure 1.20 Lattice direction in a two-dimensional lattice. Vector $\vec{a}_{\bar{1}}$ is the shortest one in this direction.

Table 1.1 Printed symbols for symmetry elements and for the corresponding symmetry operations in two dimensions

Symmetry element	Printed symbol
Reflection line (mirror line)	m
Glide line $\perp$ [01] with glide vector $1/2\vec{a}$	a or g
Glide line $\perp$ [10] with glide vector $1/2\vec{b}$	b or g
Other glide lines	g
Identity	1
n-fold rotation point, $n = 2, 3, 4, 6$	2, 3, 4, 6
Center of symmetry, inversion center	$\bar{1}$

tion $[l_1 l_2]$. Until now, we have limited our consideration of symmetry elements in two dimensions only to rotation points. However, there are still other symmetry elements in two dimensions—reflection lines (mirror lines), glide lines, or inversion centers. The last ones coincide with twofold rotation points, while the glide reflection through a line combines a reflection through this line (which is a point-group operation) with a fractional translation that is parallel to the line. In two dimensions, the allowed translations are one-half of the length of the shortest translation vector along the line. The printed symbols of the symmetry elements are collected in Table 1.1. Figures 1.21 and 1.22 show reflection lines and glide lines (both represented by their normals), respectively, in a centered rectangular lattice. The glide reflection through a line represented by its normal [01] consists of a reflection through this line and a translation by a glide vector $1/2\vec{a}$. Similarly, the glide reflection through a line represented by normal [10] consists of reflection through this line and translation by a glide vector $1/2\vec{b}$. Both those reflections are so-called axial glide reflections. Note that in two dimensions, the glide reflections are not limited only to the axial ones.

Now we will describe the international Hermann–Mauguin symbols for crystallographic space groups, known in two dimensions as *plane groups*. We will limit ourselves to the full Hermann–Mauguin symbols. Such symbol consists of a lowercase letter indicating the centering type of the conventional cell, followed by the symbol of the highest order n-fold rotation point, which is followed next by a set of symbols that correspond to the reflection or glide lines

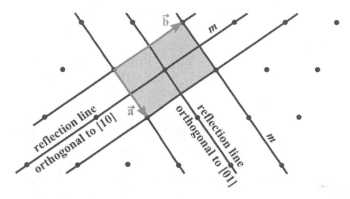

Figure 1.21 Reflection lines orthogonal to directions [10] and [01] in a centered rectangular lattice.

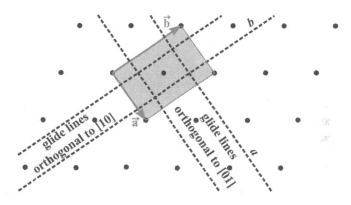

Figure 1.22 Glide lines *a* orthogonal to direction [01] and glide lines *b* orthogonal to direction [10] in a centered rectangular lattice.

present in the crystal structure. As we know already, the reflection and glide lines are represented by their normals. Table 1.2 reports such normals—lattice symmetry directions—corresponding to the secondary and tertiary positions in the set of characters indicating symmetry elements in the Hermann–Mauguin symbol. In the case of square and hexagonal lattices, instead of one symmetry direction, appears a set of symmetry directions, which are equivalent because of lattice symmetry (see Table 1.2).

The 10 crystallographic point groups and the seventeen plane groups existing in two dimensions are classified in four crystal

Table 1.2 Specification of the possible symmetry elements in the primary, secondary, and tertiary positions in the set of characters indicating symmetry elements in the Hermann–Mauguin symbol of a plane group. Mirror and glide lines are orthogonal to lattice symmetry directions specified in the last two columns

Lattice	Position of symmetry elements in the Hermann–Mauguin symbol		
	Rotation point	Symmetry directions	
	Primary	Secondary	Tertiary
Oblique	2 or 1		
Rectangular	2 or 1	[10]	[01]
Square	4	$\left\{\begin{matrix}[10]\\ [01]\end{matrix}\right\}$	$\left\{\begin{matrix}[1\bar{1}]\\ [11]\end{matrix}\right\}$
Hexagonal	6 or 3	$\left\{\begin{matrix}[10]\\ [01]\\ [\bar{1}\bar{1}]\end{matrix}\right\}$	$\left\{\begin{matrix}[1\bar{1}]\\ [12]\\ [\bar{2}\bar{1}]\end{matrix}\right\}$

systems—oblique, rectangular, square, and hexagonal—introduced already in the previous section. The largest point group belonging to a given crystal system is the point group, called *holohedry*, of a lattice belonging to this system. The rest of the groups are subgroups of the holohedry. Only those point groups belong to a given crystal system, which do not belong to the crystal systems with holohedries of fewer symmetries. Each of the 17 plane groups is the plane group of some two-dimensional crystal structure. The 10 point groups and 17 plane groups are listed in Table 1.3. Each of them belongs to one crystal system. We can observe in Table 1.3 that for some plane groups, the Hermann–Mauguin symbol has one or two characters "1" among the set of symbols that refer to the symmetry elements. Such symbol in the primary position denotes the absence of rotation points, while in the secondary or tertiary positions it signalizes the absence of reflection or glide lines. If none of the reflections through the lines represented by secondary and tertiary symmetry directions are present, then the Hermann–Mauguin symbols are reduced to two characters (a letter indicating the centering type of the conventional cell and the symbol of the rotation point). This means that the full symbol is replaced by the short Hermann–Mauguin symbol. All

Table 1.3 Classification of the crystallographic point and plane groups in crystal systems

Crystal system	Crystallographic point groups*	Crystallographic plane groups
Oblique	1, $\boxed{2}$	p1, p2
Rectangular	m, $\boxed{2mm}$	p1m1, p1g1, c1m1, p2mm, p2mg, p2gg, c2mm
Square	4, $\boxed{4mm}$	p4, p4mm, p4gm
Hexagonal	3, 6, 3m, $\boxed{6mm}$	p3, p3m1, p31m, p6, p6mm

*Symbols that are surrounded indicate point groups which are holohedries.

crystal structures from Figs. 1.6, 1.9, 1.10, and 1.12 have the same plane group $p2$, whereas the plane group of the crystal structure from Fig. 1.13 is $p1$. The last two crystal structures considered in the previous section, i.e., the honeycomb and the hexagonal structure, have the plane group $p6mm$.

We will show now a more complex two-dimensional structure on the example of a one-atom-thick layer of a boron-carbon compound (see Fig. 1.23). Its plane group has the Hermann–Mauguin symbol $p2mg$. Figure 1.23a shows a conventional unit cell of this structure. The cell contains 12 boron and 6 carbon atoms (all of them constitute the atomic basis). The origin of the conventional cell is at a twofold rotation point and on a glide line a. The two symmetry elements are specified in the primary and tertiary positions of the Hermann–Mauguin symbol, respectively, so the origin of the cell is at $21g$. In Fig. 1.23b, we show the diagram of symmetry elements, which are twofold rotation points, reflection lines $\perp[10]$, and glide lines $\perp[01]$. The group considered here is an example of a plane group whose all symmetry elements are contained in the full Hermann–Mauguin symbol. In general, a plane group may contain still additional symmetry elements.

1.5 Wyckoff Position

Wyckoff positions represent an important concept in the description of crystal structures. We will use this concept to describe the distribution of atomic basis within the unit cell. As we know already

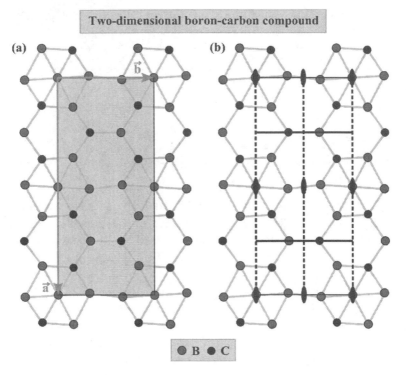

Figure 1.23 Two-dimensional structure of B_2C. A conventional unit cell (a) and the diagram of symmetry elements (b) are shown. The structure has been obtained using computational simulations. The calculated lattice parameters are $a = 10.727$ Å and $b = 4.814$ Å.

atoms may belong partially to a given unit cell and partially to the adjacent cells. Also lattice points placed at the vertices of the unit cell belong partially to it. The same may happen to other points of interest, for instance, the points in which are placed symmetry elements of the crystal structure. It is convenient, however, to consider entire atoms (points) instead of its fractions. This is schematically shown in Fig. 1.24. The entire atom (point) has the location as close as possible to the cell origin (see Figs. 1.24a and 1.24b to the right). In these figures, the coordinates of an atom (point) are expressed in units of a and b (lengths of basis vectors) and given by a doublet x, y. In the right part of Fig. 1.24, we

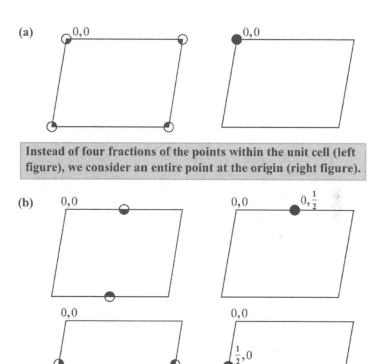

Figure 1.24 Atoms (points) in different positions within the unit cell of the oblique lattice. (a) Each of the atoms (points) located at the vertices contribute with a fraction to the unit cell (draw to the left). The sum of these fractions represents an entire atom (point), which is located at the origin of the cell (draw to the right). (b) In each case shown to the right, an entire atom (point) represents the sum of two fractions of atoms (points) shown to the left, and is located at the position which is closer to the origin (draw to the right). The basis vector $\vec{a}$ is pointing downwards and vector $\vec{b}$ to the right. The coordinates of the entire atoms (points) are given in units of a and b.

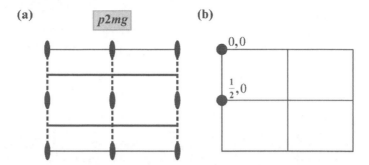

Figure 1.25 (a) Diagram of symmetry elements for the *p2mg* plane group. (b) An example of symmetrically equivalent points. The basis vector $\vec{a}$ is pointing downwards and vector $\bar{b}$ to the right. The coordinates of points are given in units of *a* and *b*.

have atoms (points) of following coordinates: $0, 0$; $0, \frac{1}{2}$; and $\frac{1}{2}, 0$ in Figs. 1.24a, 1.24b (top), and 1.24b (bottom), respectively.

Turning now to the symmetrically equivalent points, we give an example of such points in Fig. 1.25. In Fig. 1.25a, we have the diagram of symmetry elements for the plane group *p2mg* and in Fig. 1.25b two symmetrically equivalent points are shown. One point may be obtained from the other by applying a plane group symmetry operation, which in this case is the reflection through the upper reflection line shown in the diagram of symmetry elements.

As next, we will introduce the concept of Wyckoff positions.

- A *general position* is a set of symmetrically equivalent points within the unit cell, such that each of them is located out of any symmetry element of the plane group.
- A *special position* is a set of symmetrically equivalent points within the unit cell, such that each of them is located on at least one symmetry element of the plane group.

In both cases, the set of points may be substituted by an equivalent set of points, in which some points are located closer to the origin of the unit cell by applying integer lattice translations. The general position is only one for each plane group, while there are in general few special positions. Figure 1.26

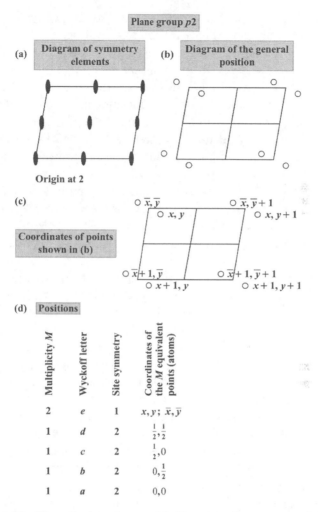

Figure 1.26 The *p*2 plane group. (a) Diagram of symmetry elements. (b) Diagram of the general position. (c) Coordinates of points shown in (b). (d) Wyckoff positions. In (a), (b), and (c) the basis vector $\vec{a}$ is pointing downwards and vector $\vec{b}$ to the right.

provides the information about the general position and special positions for the *p*2 plane group. In this figure, we have in (a) the diagram of symmetry elements, which shows the relative location of the symmetry elements. In (b) is given the diagram of the general position. This diagram illustrates the arrangement of a set of symmetrically equivalent points of general position. Finally, in (c) are given the coordinates of points shown in (b). Additionally, in Fig. 1.26d, the Wyckoff positions are given. Open circles in the diagram of the general position (see Fig. 1.26b) give the positions of 8 points obtained from a point of coordinates x, y (located out of any symmetry element) under the action of the $p2$ group symmetry operations: identity operation at first, followed by translations by lattice translation vectors $\vec{a}$, $\vec{b}$, and $\vec{a} + \vec{b}$ and after that by the rotations about twofold rotation points located at the vertices of the cell. The coordinates of these 8 points are listed in Fig. 1.26c. We can observe in this figure that only two of them, of coordinates x, y and $\bar{x} + 1, \bar{y} + 1$ (where $-x$ and $-y$ are substituted by symbols $\bar{x}$ and $\bar{y}$, respectively), are located inside the unit cell. Note that at the Wyckoff general position, instead of $\bar{x} + 1, \bar{y} + 1$, the doublet $\bar{x}, \bar{y}$ is rather used, since the coordinates are formulated modulo 1. At the bottom of Fig. 1.26, we can see the block under the title "Positions" that is organized from bottom to top, and provides the information about all special positions and, at the top, about the general position. This block is organized in four columns. In the first column, the multiplicity M of the Wyckoff position is given. This is the number of symmetrically equivalent points per unit cell that correspond to that position. In the second column, we have the Wyckoff letter, which is assigned to each Wyckoff position of a given plane group in the way shown in Fig. 1.26. The letters start with a at the bottom of the block and continue upwards in alphabetical order up to the letter that corresponds to the general position. In the third column of the block, the *site symmetries* of Wyckoff positions are given. The site symmetry group collects all operations that leave each of the points belonging to the considered position invariant. In the case of the general position, independently of the plane group, the site symmetry group is limited to the identity only, while in case of a special position, the site symmetry group has to contain at least one symmetry operation in addition to identity. In our example of the $p2$ plane group, the site symmetry groups for all special positions (a, b, c, and d) are identical and contain rotations around twofold rotation points. The last column of the block gives the coordinate doublets for the M symmetrically equivalent points belonging to a given position.

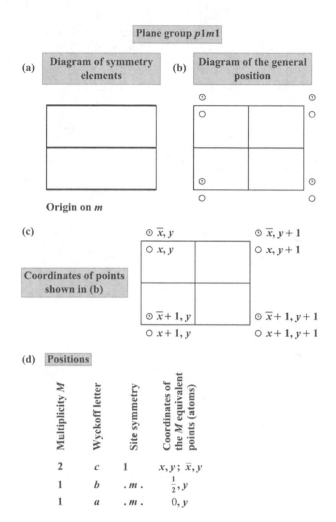

Figure 1.27 The *p*1*m*1 plane group. (a) Diagram of symmetry elements. (b) Diagram of the general position. (c) Coordinates of points shown in (b). (d) Wyckoff positions. In (a), (b), and (c), the basis vector $\vec{a}$ is pointing downwards and the vector $\vec{b}$ to the right.

Figure 1.27 provides the description of the *p*1*m*1 plane group. We can observe in Fig. 1.27b that in this case, like in the case of the *p*2 plane group, the unit cell contains only two of the eight symmetrically equivalent points shown in the diagram of the general

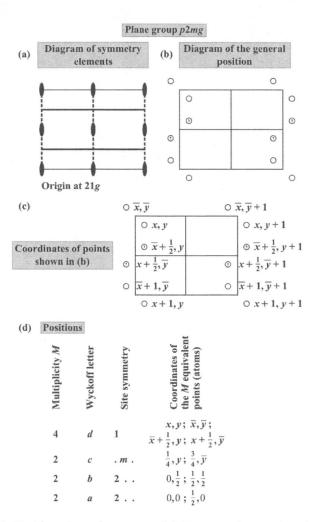

Plane group p2mg

(a) Diagram of symmetry elements

(b) Diagram of the general position

Origin at 21g

(c) Coordinates of points shown in (b)

$\bar{x},\bar{y}$ $\bar{x},\bar{y}+1$
x,y $x,y+1$
$\bar{x}+\frac{1}{2},y$ $\bar{x}+\frac{1}{2},y+1$
$x+\frac{1}{2},\bar{y}$ $x+\frac{1}{2},\bar{y}+1$
$\bar{x}+1,\bar{y}$ $\bar{x}+1,\bar{y}+1$
$x+1,y$ $x+1,y+1$

(d) **Positions**

Multiplicity M	Wyckoff letter	Site symmetry	Coordinates of the M equivalent points (atoms)
4	d	1	$x,y;\ \bar{x},\bar{y};$ $\bar{x}+\frac{1}{2},y;\ x+\frac{1}{2},\bar{y}$
2	c	. m .	$\frac{1}{4},y;\ \frac{3}{4},\bar{y}$
2	b	2 . .	$0,\frac{1}{2};\ \frac{1}{2},\frac{1}{2}$
2	a	2 . .	$0,0;\ \frac{1}{2},0$

Figure 1.28 The *p2mg* plane group. (a) Diagram of symmetry elements. (b) Diagram of the general position. (c) Coordinates of points shown in (b). (d) Wyckoff positions. In (a), (b), and (c) the basis vector $\vec{a}$ is pointing downwards and vector $\vec{b}$ to the right.

position. Their coordinates are x,y and $\bar{x}+1,y$, but at the Wyckoff general position c, instead of $\bar{x}+1,y$, the doublet $\bar{x},y$ is used. Because of the presence of mirror lines, the points shown in the diagram of general position are represented now using both open circles and circles containing a comma inside, to signalize the change of orientation of an object after reflection.

As a last example, in Fig. 1.28 is given the description of the *p2mg* plane group. In this case, the Wyckoff general position *d* contains four equivalent points, and each of the three special positions *a*, *b*, and *c* contains two equivalent points. In the case of the special positions *a* and *b*, this is due to the presence of the upper mirror line, while in the case of the special position *c* due to the presence of the twofold rotation point at the center of the cell. The last operation generates a point of coordinates $\frac{3}{4}, \bar{y} + 1$ from a point of coordinates $\frac{1}{4}, y$ (at the Wyckoff special position *c* instead of $\frac{3}{4}, \bar{y} + 1$, the doubled $\frac{3}{4}, \bar{y}$ is used).

Problems

Exercise 1.1   Figure 1.29 shows a one-dimensional crystal structure composed of three different atoms. Propose a lattice for this structure and comment your choice. Draw the unit cell. Is this cell a conventional cell?

Figure 1.29   One-dimensional crystal structure composed of three types of atoms.

Exercise 1.2   In one dimension, all lattices are primitive. The centering type of the conventional primitive cell is denoted by the script letter $\wp$. There are only two types of one-dimensional point groups: 1 and $m \equiv \bar{1}$. At the same time, there are only two types of one-dimensional space groups (so-called *line groups*). Their Hermann-Mauguin symbols are $\wp 1$ and $\wp m \equiv \wp \bar{1}$. The group $\wp 1$ consists of translations only, whereas $\wp m$ possesses also reflections through points. Specify the line groups of the crystal structures shown in Figs. 1.1 and 1.29. Justify your answer.

Exercise 1.3   Draw two examples of conventional unit cells for the crystal structure defined in Fig. 1.13, different from the one proposed in this figure.

Exercise 1.4   Figure 1.30 shows a hexagonal lattice. What lattice will be obtained if we place an additional point in the geometric center of each equilateral triangle in Fig. 1.30? Draw the basis vectors of the new lattice.

Exercise 1.5   Consider the hexagonal and centered rectangular lattices shown in Figs. 1.30 and 1.31, respectively.

(a) For the hexagonal lattice, show the rotation points of order two by placing, within the conventional cell, the appropriate graphical symbol.
(b) Repeat (a) for the centered rectangular lattice.

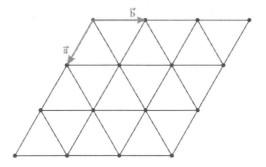

Figure 1.30   A hexagonal lattice.

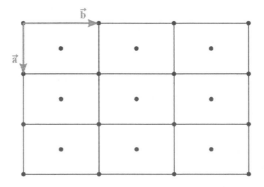

Figure 1.31   A centered rectangular lattice.

Exercise 1.6   Consider the lattice from Fig. 1.30.

(a) What crystal structure will be obtained if we attach to each lattice point a basis that has two identical atoms in the positions given by vectors $\vec{r}_1 = 2/3\vec{a} + 1/3\vec{b}$ and $\vec{r}_2 = 1/3\vec{a} + 2/3\vec{b}$? Draw this structure.

(b) If, instead of using identical atoms, we use in (a) a basis consisting of one boron and one nitrogen atom, then the resulting structure will be an isolated atomic sheet of the hexagonal phase of boron nitride (α-BN). What are the orders of rotation points in the two-dimensional boron nitride structure? Draw this structure and show the rotation points.

Exercise 1.7 In Fig. 1.32a, we show a hexagon that has the point symmetry of an infinite two-dimensional crystal structure composed of two types of atoms. Show that this structure may be obtained by attaching to each lattice point of the hexagonal lattice from Fig. 1.30 a basis that has a larger atom at the position given by vector $\vec{r}_1 = \vec{0}$ and two smaller atoms at the positions given by vectors $\vec{r}_2 = 2/3\vec{a} + 1/3\vec{b}$ and $\vec{r}_3 = 1/3\vec{a} + 2/3\vec{b}$. The superposition of what substructures can reproduce the structure in consideration?

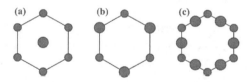

Figure 1.32 Hexagons that have the same point symmetry as the two-dimensional infinite structures composed of two types of atoms.

Exercise 1.8 Show graphically that the honeycomb structure (see Fig. 1.14) is nothing more than the superposition of two identical hexagonal substructures shifted one with respect to the other.

Exercise 1.9 Figure 1.33 shows a unit cell for a two-dimensional lattice.

(a) What type of lattice is this?
(b) Draw the primitive unit cell for this lattice as shown in Fig. 1.19.

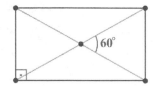

Figure 1.33   Unit cell for a two-dimensional lattice.

Exercise 1.10   In a square lattice, draw the lattice directions denoted by $[1\bar{1}]$ and $[11]$. In a second draw, show the reflection lines represented by those directions.

Exercise 1.11   Repeat Exercise 1.10 for the case of a hexagonal lattice and lattice directions denoted by

(a) $[10]$, $[01]$, and $[\bar{1}\bar{1}]$.
(b) $[1\bar{1}]$, $[12]$, and $[\bar{2}\bar{1}]$.

Exercise 1.12   Figure 1.32b shows a hexagon that has the point symmetry of an infinite two-dimensional structure of α-BN (for reference, see also Exercise 1.6).

(a) Draw this structure locating the hexagon from Fig. 1.32b in each point of the lattice from Fig. 1.30 in such a way that its center coincides with a lattice point.
(b) The Hermann–Mauguin symbol for the plane group of the α-BN structure is *p3m1*. Show on the draw from (a) the symmetry elements indicated in the different positions of the Hermann–Mauguin symbol.

Exercise 1.13   Repeat Exercise 1.12 for the structure defined in Fig. 1.32c. The Hermann–Mauguin symbol for the plane group of this structure is *p6mm*.

Exercise 1.14   The Hermann–Mauguin symbols for the plane groups of rectangular and centered rectangular lattices are *p2mm* and *c2mm*, respectively.

(a) Draw a rectangular lattice and show the diagram of symmetry elements of its plane group.

(b) Place in the draw of the centered rectangular lattice shown in Fig. 1.31 the diagram of symmetry elements of its plane group. Note, that the diagram of symmetry elements should contain all types of symmetry elements, not only those indicated in the Hermann–Mauguin symbol $c2mm$.

Exercise 1.15 The point groups of the unit cells shown in Fig. 1.19 are at the same time the point groups (holohedries) of the corresponding lattices, with the exception of the hexagonal lattice.

(a) Figure 1.19b shows a unit cell for the rectangular lattice and the twofold rotation point at the center of the cell. Place the rest of the point symmetry elements of the cell and specify the Hermann–Mauguin symbol for the point group of the rectangular lattice.
(b) Repeat (a) for a centered rectangular lattice.

Exercise 1.16 In Fig. 1.34, we show a two-dimensional boron-carbon compound of B_3C stoichiometry. How many boron atoms and how many carbon atoms do belong to the unit cell defined in the figure? Justify your answer.

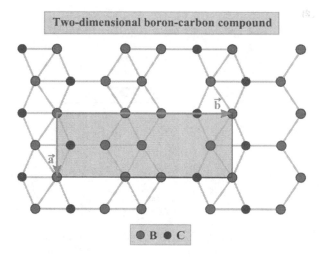

Figure 1.34 Two-dimensional B_3C structure. A conventional unit cell is shown. The structure has been obtained using computational simulations. The calculated lattice parameters are $a = 2.827$ Å and $b = 7.750$ Å.

Exercise 1.17   Figure 1.34 shows a conventional unit cell for a two-dimensional B_3C structure.

(a) What symmetry elements do cross the cell origin?
(b) Determine the Hermann–Mauguin symbol for the plane group of this structure and show the diagram of symmetry elements of the plane group.

Exercise 1.18   The Hermann–Mauguin symbol for the plane group of the square lattice is *p4mm*.

(a) Draw a square lattice and show the diagram of symmetry elements of its plane group.
(b) Which symmetry elements are not indicated in the Hermann-Mauguin symbol?

Exercise 1.19   The (001) surface of the NaCl crystal structure is one of the most studied insulating surfaces due to its simple structure, stable stoichiometry, and easy preparation. The ideal surface structures for many materials can be easily deduced from their bulk structures. For example, the (001) surface of NaCl that would be expected from the bulk structure is shown in Fig. 1.35.

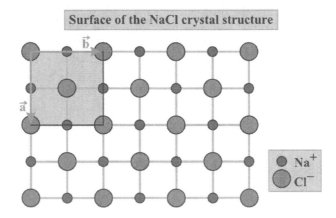

Figure 1.35   The ideal (001) surface of the NaCl crystal structure. Note that usually atoms at the crystal surfaces assume a different structure than that of the bulk atoms; however, this was not taken into account in this figure. Vectors $\vec{a}$ and $\vec{b}$ are primitive translation vectors of the three-dimensional NaCl structure. The experimental lattice constant is $a = b = 5.6401$ Å.

(a) Draw a conventional unit cell for this surface.

(b) Repeat (a) and (b) from Exercise 1.17 for this surface.

Exercise 1.20 Figure 1.36 shows an example of a two-dimensional boron structure with its conventional cell which contains four atoms. In this figure, there are also drawn one reflection line and one glide

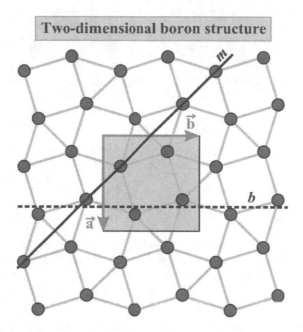

Figure 1.36 A two-dimensional boron structure with its conventional unit cell. One reflection line and one glide line are also specified in the draw. The structure has been obtained using computational simulations. The calculated lattice parameters are $a = b = 3.228$ Å.

line. Take one atom from the draw and demonstrate that the glide reflection through this line changes the position of this atom such that it coincides with some other atom.

Exercise 1.21 Determine the plane group of the two-dimensional boron structure shown in Fig. 1.36, and draw the diagram of its symmetry elements (two of them are already shown in the figure). Specify the symmetry elements that are not indicated in the

Hermann–Mauguin symbol. What symmetry elements are located at (or cross) the cell origin?

Exercise 1.22 Figure 1.37 shows a two-dimensional boron structure.

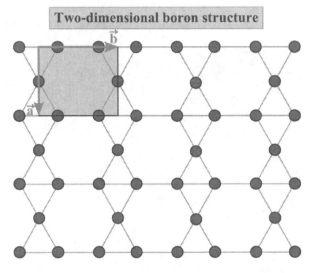

Figure 1.37 Another example (see Fig. 1.36) of a two-dimensional boron structure. The structure has been obtained using computational simulations. The calculated lattice parameters are $a = 2.882$ Å and $b = 3.328$ Å.

(a) How many atoms do constitute the atomic basis?
(b) Determine the plane group of the structure and show the diagram of its symmetry elements.
(c) What symmetry elements do cross the cell origin? Could the conventional origin have a different position with respect to the boron atoms? Justify your answer.

Exercise 1.23 Inside the unit cell of the two-dimensional B_2C structure shown in Fig. 1.23a, place the appropriate labels on carbon atoms in different Wyckoff positions specified in Table 1.4. The Wyckoff positions for the $p2mg$ plane group, which is the group of the B_2C structure, are described in Fig. 1.28.

Exercise 1.24 Repeat Exercise 1.23 for boron atoms.

Exercise 1.25 Figure 1.34 shows a two-dimensional B_3C structure. The coordinates of 6 boron atoms that lie within the unit cell, given in units of a and b, are $0, 0$; $0, 0.390$; $0, 0.602$; $\frac{1}{2}, 0.273$; $\frac{1}{2}, 0.484$; and $\frac{1}{2}, 0.874$. Whereas, the coordinates of 2 carbon atoms (within the unit cell) are $0, 0.795$ and $\frac{1}{2}, 0.080$.

(a) What is the plane group of this structure?
(b) Specify the Wyckoff positions of B and C atoms located within the unit cell. Organize this information in a way shown in Table 1.4. Also place, inside the unit cell from Fig. 1.34, the appropriate labels on boron and carbon atoms.

Table 1.4 Wyckoff positions and the coordinates of the corresponding boron (first block) and carbon (second block) atoms located within the unit cell shown in Fig. 1.23a. The coordinates are given in units of a and b

Atom	Wyckoff position	Coordinates of equivalent positions*
B1	2a	$0, 0; \frac{1}{2}, 0$
B2	2c	$\frac{1}{4}, 0.658; \frac{3}{4}, 0.342$
B3	4d	$x_1, y_1; \bar{x}_1 + 1, \bar{y}_1 + 1; \bar{x}_1 + \frac{1}{2}, y_1; x_1 + \frac{1}{2}, \bar{y}_1 + 1$
B4	4d	$x_2 + 1, y_2; \bar{x}_2, \bar{y}_2 + 1; \bar{x}_2 + \frac{1}{2}, y_2; x_2 + \frac{1}{2}, \bar{y}_2 + 1$
C1	2c	$\frac{1}{4}, 0.343; \frac{3}{4}, 0.657$
C2	4d	$x, y + 1; \bar{x} + 1, \bar{y}; \bar{x} + \frac{1}{2}, y + 1; x + \frac{1}{2}, \bar{y}$

*$x = 0.128$, $y = -0.162$; $x_1 = 0.137$, $y_1 = 0.162$; $x_2 = -0.009$, $y_2 = 0.335$

Exercise 1.26 In Table 1.5 are given the Wyckoff positions and the coordinates (expressed in units of the length of surface basis vectors) for the basis atoms of the NaCl(001) surface shown in Fig. 1.35 (plane group $p4mm$).

(a) Draw a conventional unit cell for the NaCl(001) surface (choose the origin of a conventional unit cell in accord with the data provided in Table 1.5).

Table 1.5 Wyckoff positions and the corresponding coordinates of Na and Cl ions located within the unit cell of the NaCl(001) surface (plane group *p4mm*). The coordinates are given in units of the length of surface basis vectors

Ion	Wyckoff position	Coordinates of equivalent positions
Na^+	1*b*	$\frac{1}{2}, \frac{1}{2}$
Cl^-	1*a*	$0, 0$

(b) What is the site symmetry for the Wyckoff special position 1*a*? Justify your answer.

(c) Repeat (b) for the special position 1*b*.

Exercise 1.27 The basis atoms of the two-dimensional boron structure shown in Fig. 1.36 constitute the special Wyckoff position 4*c* of the plane group *p4gm*, in which the atoms lie on the reflection lines $m\perp[1\bar{1}]$ and $m\perp[11]$. The coordinates of the basis atoms of the structure shown in Fig. 1.36 are given in Table 1.6.

(a) Associate the coordinates given in the Table 1.6 with atoms located within the unit cell shown in Fig. 1.36.

Table 1.6 Wyckoff position and the corresponding coordinates of boron atoms located within the unit cell shown in Fig. 1.36 (plane group *p4gm*). The coordinates are given in units of *a*

Atom	Wyckoff position	Coordinates of equivalent positions
B	4*c*	0.173, 0.673; 0.827, 0.327; 0.327, 0.173; 0.673, 0.827

(b) The coordinates of atoms in the Wyckoff special position 4*c* are $x, x + \frac{1}{2}$; $\bar{x}, \bar{x} + \frac{1}{2}$; $\bar{x} + \frac{1}{2}, x$; and $x + \frac{1}{2}, \bar{x}$. Using the data given in Table 1.6, verify that the basis atoms of the boron structure shown in Fig. 1.36 constitute indeed the special Wyckoff position 4*c*.

(c) What is the site symmetry for the special position 4*c*? Justify your answer.

Table 1.7 Wyckoff positions and the corresponding coordinates of boron atoms located within the unit cell shown Fig. 1.37 (plane group $p2mm$). The coordinates are given in units of a and b

Atom	Wyckoff position	Coordinates of equivalent positions
B1	1c	$\frac{1}{2}, 0$
B2	2g	$0, \frac{1}{4}; \ 0, \frac{3}{4}$

Exercise 1.28 In Table 1.7 are given the Wyckoff positions and the corresponding coordinates of atoms, which represent the basis for the two-dimensional boron structure shown in Fig. 1.37 (plane group $p2mm$).

(a) Identify atoms B1 and B2 within the unit cell.
(b) What are the site symmetries for the special positions 1c and 2g?

Exercise 1.29 The line group p-m consists of translations in one dimension and reflections through points. Give the Wyckoff positions for the line group p-m in the way done for the plane groups in Figs. 1.26–1.28.

Exercise 1.30 Figure 1.38 shows a one-dimensional structure composed of two types of atoms labeled A and B. The coordinates, expressed in units of a, of atom A and two atoms B that lie within the unit cell are 0, $\frac{1}{4}$, and $\frac{3}{4}$, respectively.

(a) What is the line group of the structure in consideration?
(b) Specify the Wyckoff positions of atoms located within the unit cell.

A   B   B

Figure 1.38 One-dimensional crystal structure composed of two types of atoms, labeled A and B. A conventional unit cell is also shown.

Chapter 2

Three-Dimensional Crystal Lattices

2.1 Introduction

In the case of a three-dimensional lattice, a primitive unit cell has the shape of a parallelepiped defined by three non-collinear and not all in the same plane primitive translation vectors $\vec{a}_1$, $\vec{a}_2$, and $\vec{a}_3$. The most general example of a primitive unit cell is shown in Fig. 2.1.

The translation symmetry of a two- or three-dimensional lattice imposes certain restrictions on its point symmetry elements, as shown in the previous chapter for the case of a two-dimensional lattice. The allowed orders of symmetry axes in the three-dimensional lattice are the same as the orders of symmetry points in two dimensions: 1, 2, 3, 4, and 6. As a consequence, in two and three dimensions, only certain lattice types are possible. In order to find them in three dimensions, we will proceed in a similar way as it was done for the two-dimensional case. First, we will consider certain finite three-dimensional figures whose symmetry axes are of the orders that are allowed in the lattices.

Basic Elements of Crystallography (2nd Edition)
Nevill Gonzalez Szwacki and Teresa Szwacka
Copyright © 2016 Pan Stanford Publishing Pte. Ltd.
ISBN 978-981-4613-57-6 (Hardcover), 978-981-4613-58-3 (eBook)
www.panstanford.com

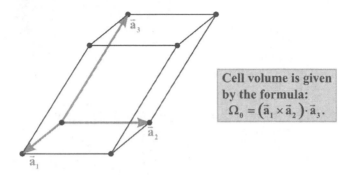

Cell volume is given
by the formula:
$$\Omega_0 = \left(\vec{a}_1 \times \vec{a}_2\right)\cdot \vec{a}_3.$$

Figure 2.1   A primitive unit cell of a three-dimensional lattice.

2.2 Examples of Symmetry Axes of Three-Dimensional Figures

An object which has one or more symmetry axes of orders 1, 2, 3, 4, or 6 may have the shape of such a solid figure as parallelepiped, regular tetrahedron or octahedron, or hexagonal prism. When the point symmetry of a lattice is such that the highest order of the *n*-fold symmetry axis is only one, a parallelepiped of the lowest possible symmetry (see Fig. 2.1) represents a solid figure that has the same point symmetry as the lattice. In Fig. 2.2, we show other parallelepipeds whose shapes allow for the presence of two- and (or) fourfold symmetry axes. We can see in each case the rotation axes that cross the geometric centers of the parallelepiped faces. In the case of a square prism (Fig. 2.2b) and a cube (Fig. 2.2c), there are also present other rotation axes.

Unit cells, with the same symmetry as the point symmetry of an important number of lattices, have shapes of the three parallelepipeds shown in Fig. 2.2. The symmetry center of each parallelepiped overlaps with its geometric center. This is a common property of all point symmetry elements. Obviously, the orders of rotation axes and the number of axes of the same order depend on the shape of the parallelepiped. For example, a cube (Fig. 2.2c) has three fourfold rotation axes. Each of them is defined by the geometric centers of two square faces parallel to each other. The cube has a total of 13 rotation axes—besides the three fourfold axes

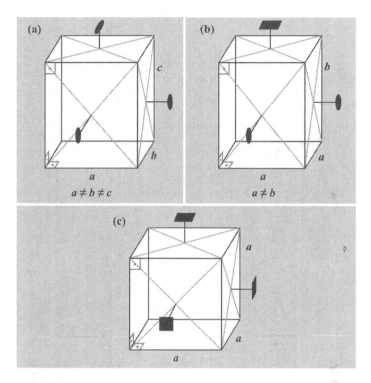

Figure 2.2 Some rotation axes of three solid figures: (a) rectangular prism, (b) square prism, and (c) cube.

shown in Fig. 2.2c it has two- and threefold axes. The case of the cube will be considered in more details in the next section.

As we can see in Fig. 2.3a,b, a regular tetrahedron and a regular octahedron, respectively, can be inscribed in a cube. A tetrahedron has three mutually perpendicular twofold rotation axes instead of the fourfold rotation axes of the cube (see Fig. 2.3a). Each of them is defined by the centers of its two edges. A tetrahedron does not represent a unit cell of any lattice, but it is relevant in the description of important crystal structures (especially in the description of their symmetry). Contrary to the tetrahedron, a regular octahedron has the same three mutually perpendicular fourfold rotation axes as the cube has (see Fig. 2.3b) with the difference that in the case of an octahedron a fourfold axis is defined by two vertices and in the case of the cube by the geometric centers of two faces (the number of

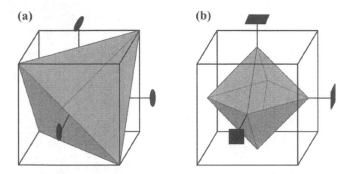

Figure 2.3 A regular tetrahedron (a) and a regular octahedron (b) inscribed in a cube.

octahedron vertices coincides with the number of cube faces and *vice versa*).

The solid figure which has a sixfold rotation axis takes on the shape of a regular hexagonal prism. Therefore, the unit cell of the same point symmetry as that of a hexagonal lattice in three

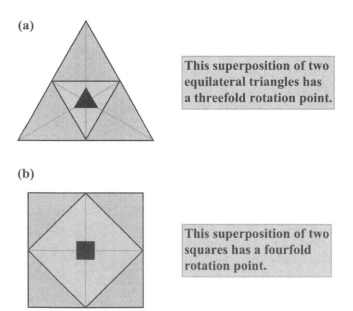

(a) This superposition of two equilateral triangles has a threefold rotation point.

(b) This superposition of two squares has a fourfold rotation point.

Figure 2.4 Symmetry points of a superposition of plane figures: (a) two equilateral triangles and (b) two squares.

dimensions (the so-called *Wigner–Seitz cell*) has the shape of a regular hexagonal prism. This will be considered in more details later.

Before continuing with the three-dimensional case, we will look shortly at the symmetry points of a superposition of plane figures. The superposition of two equilateral triangles with a common geometric center has a threefold rotation point. This is shown in Fig. 2.4a. A similar superposition of two squares has a fourfold rotation point (see Fig. 2.4b). Both examples will be helpful in further considerations of rotation axes in some three-dimensional lattices.

2.3 Rotation Axes of a Cube

Let us now continue with the consideration of the possible rotation axes of a cube. First, we will look at the twofold rotation axes. Each of them is defined by the centers of two edges, as shown in Fig. 2.5. So, a cube has a total of 6 twofold rotation axes.

It is easy to demonstrate that the body diagonals of a cube represent its threefold rotation axes. We can see in Fig. 2.6 that the shown body diagonal connects two opposite cube vertices. The remaining 6 vertices form two groups, with 3 vertices each, which represent the vertices of two equilateral triangles. Each of

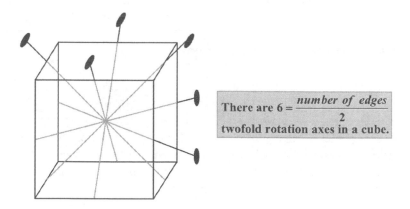

There are $6 = \dfrac{number\ of\ edges}{2}$ twofold rotation axes in a cube.

Figure 2.5   Six twofold rotation axes of a cube.

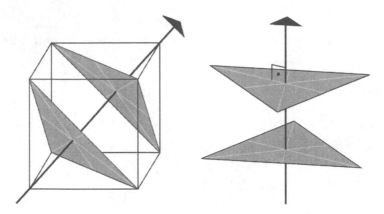

Figure 2.6 Each diagonal of a cube represents one of its threefold rotation axis.

the triangles is lying in a plane orthogonal to the diagonal and its geometric center overlaps with the point where the diagonal intersects the plane of the triangle. It is obvious that after rotating the cube by an angle $2\pi/3$ (or its multiples), the new positions of the cube vertices (those out of the axis) overlap with some "old" positions of the vertices. Therefore, this transformation leaves the cube invariant. Besides the axis shown in Fig. 2.6, there are 3 more threefold rotation axes in the cube, that is, as many as the number of body diagonals. In conclusion, a cube has a total of 13 rotation axes. All of them are shown in Fig. 2.7.

2.4 Rotation Axes of a Set of Points

Now, we will concentrate our attention on a system consisting of a set of 8 points (or atoms) located at the vertices of a cube. The symmetry axes of this set of points are the same as the symmetry axes of the cube. If we add one additional point in the middle of the cube, then the symmetry of the resulting system will remain the same, since this point will be a common point of all the axes and also other symmetry elements. Also, if we add points in the middle of the faces of the cube, then the symmetry of this new 14-point system (shown in Fig. 2.8) will still remain the same as in the system

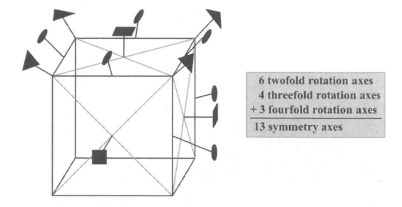

Figure 2.7   The 13 rotation axes of a cube.

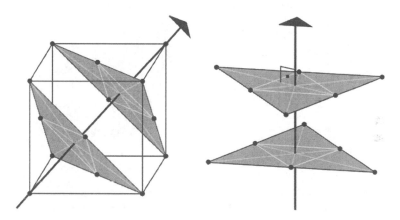

Figure 2.8   The system of 14 points placed at the vertices and in the geometric centers of the faces of a cube have the same threefold rotation axes as the cube.

consisting of only 8 points. For example, it is easy to see, comparing Figs. 2.6 and 2.8, that the threefold rotation axes are present in this 14-point system. The six new points will form two groups of three points each, which are located in the middle of triangle edges, as appears in Fig. 2.8. The axial view of one of the triangles from Fig. 2.8 is shown in Fig. 2.9.

Let us now consider the fourfold rotation axes in the case of the 14-point system in consideration. We can observe in Fig. 2.10 that

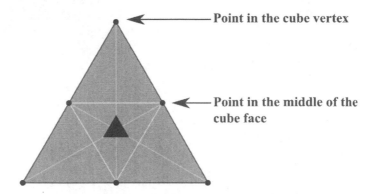

Point in the cube vertex

Point in the middle of the cube face

Figure 2.9 Axial view of one of the triangles from Fig. 2.8.

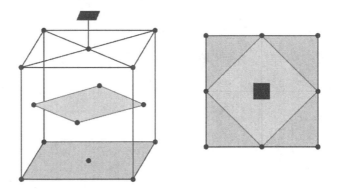

Figure 2.10 Fourfold rotation axis of a system consisting of 14 points located at the vertices and centers of the faces of a cube. On the right is shown the projection of the 14 points on the cube bottom face.

of a total of six points, in the middle of the cube faces, two are on the axis and the remaining four represent vertices of a square lying in a plane orthogonal to the axis. If we project the 14 points on a plane orthogonal to the axis, then we will obtain a superposition of two squares shown on the right of Fig. 2.10. Thus, we can say that the 14-point system has the same three fourfold rotation axes as the cube. Besides that, the system of points has six twofold rotation axes. Finally, we can conclude that in the three cases described above, and shown in Fig. 2.11, we have the same 13 symmetry axes as were identified before in a cube.

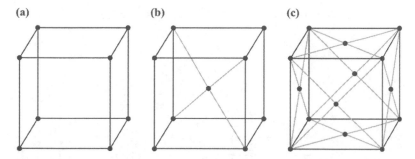

Figure 2.11 Three systems consisting of (a) 8 points at the vertices of a cube, (b) 9 points at the vertices and the geometric center of a cube, and (c) 14 points at the vertices and face centers of a cube. Each set of points has the same 13 rotation axes as the cube.

2.5 Crystal Systems

In three dimensions, the space and point groups of crystal structures are classified in 7 crystal systems: triclinic, monoclinic, orthorhombic, tetragonal, trigonal, hexagonal, and cubic. The largest crystallographic point group, called holohedry, belonging to a given crystal system is the point group of lattices belonging to this system. The holohedries coincide with the point groups of the solid figures shown in Fig. 2.12. Each solid figure (or strictly speaking its point group) represents one crystal system. For each case shown in Fig. 2.12, the highest-order rotation axis is given. If there is more than one such axis, it is also shown in the figure.

The solid figures shown in Fig. 2.12 are defined by the basis vectors $\vec{a}$, $\vec{b}$, and $\vec{c}$ parallel (in most cases) to the main symmetry axes, if there are any in the lattices belonging to the crystal system (represented by a solid figure). In the lattices which belong to the triclinic system, there are no symmetry axes at all or, more precisely, there are only onefold axes. Thus, no basis vector is fixed by symmetry (see Fig. 2.12a) and $\vec{a}$, $\vec{b}$, and $\vec{c}$ are just three non-collinear and not all in the same plane primitive translation vectors of a triclinic lattice. As we know already, the conventional basis vectors should be chosen to be the shortest possible and the angles between them should be as close to 90° as possible but greater than or equal to 90°.

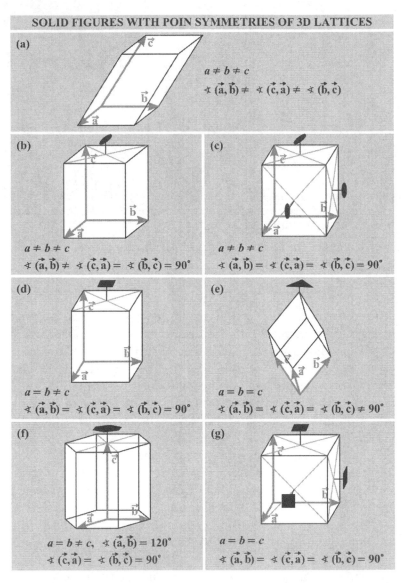

SOLID FIGURES WITH POIN SYMMETRIES OF 3D LATTICES

(a)
$$a \neq b \neq c$$
$$\sphericalangle\,(\vec{a},\vec{b}) \neq \;\sphericalangle\,(\vec{c},\vec{a}) \neq \;\sphericalangle\,(\vec{b},\vec{c})$$

(b)
$$a \neq b \neq c$$
$$\sphericalangle\,(\vec{a},\vec{b}) \neq \;\sphericalangle\,(\vec{c},\vec{a}) = \;\sphericalangle\,(\vec{b},\vec{c}) = 90°$$

(c)
$$a \neq b \neq c$$
$$\sphericalangle\,(\vec{a},\vec{b}) = \;\sphericalangle\,(\vec{c},\vec{a}) = \;\sphericalangle\,(\vec{b},\vec{c}) = 90°$$

(d)
$$a = b \neq c$$
$$\sphericalangle\,(\vec{a},\vec{b}) = \;\sphericalangle\,(\vec{c},\vec{a}) = \;\sphericalangle\,(\vec{b},\vec{c}) = 90°$$

(e)
$$a = b = c$$
$$\sphericalangle\,(\vec{a},\vec{b}) = \;\sphericalangle\,(\vec{c},\vec{a}) = \;\sphericalangle\,(\vec{b},\vec{c}) \neq 90°$$

(f)
$$a = b \neq c, \;\; \sphericalangle\,(\vec{a},\vec{b}) = 120°$$
$$\sphericalangle\,(\vec{c},\vec{a}) = \;\sphericalangle\,(\vec{b},\vec{c}) = 90°$$

(g)
$$a = b = c$$
$$\sphericalangle\,(\vec{a},\vec{b}) = \;\sphericalangle\,(\vec{c},\vec{a}) = \;\sphericalangle\,(\vec{b},\vec{c}) = 90°$$

Figure 2.12 Solid figures with point symmetries of lattices belonging to the 7 crystal systems in three dimensions: (a) triclinic, (b) monoclinic, (c) orthorhombic, (d) tetragonal, (e) trigonal, (f) hexagonal, and (g) cubic. The solid figures are defined by the basis vectors $\vec{a}$, $\vec{b}$, and $\vec{c}$ parallel (in most cases) to the main symmetry axes, if there are any in a lattice belonging to a given crystal system.

In the case of lattices belonging to the monoclinic system, only one symmetry direction exists. This is shown in Fig. 2.12b, where for the solid figure defined by the basis vectors of a monoclinic lattice, it is highlighted one twofold rotation axis (parallel to the basis vector $\vec{c}$). The restriction $\sphericalangle(\vec{c}, \vec{a}) = \sphericalangle(\vec{b}, \vec{c}) = 90°$ guarantees the presence of this unique symmetry direction.

In the lattices belonging to the orthorhombic, tetragonal, and cubic systems three mutually perpendicular symmetry directions coexist and the basis vectors $\vec{a}$, $\vec{b}$, and $\vec{c}$, which define the solid figures shown in Figs. 2.12c,d,g, are parallel to them. Figures 2.12c,g show the three twofold and fourfold rotation axes present in lattices belonging to the orthorhombic and cubic systems, respectively, while Fig. 2.12d shows the fourfold rotation axis present in lattices of the tetragonal system. The remaining rotation axes (not shown in Fig. 2.12d) are twofold.

The lattices that have only one threefold rotation axis belong to the trigonal crystal system, while the lattices that have only one sixfold rotation axis belong to the hexagonal and trigonal crystal systems (the threefold rotation axis represents a sub-element of the sixfold rotation axis). A solid figure that possesses a sixfold rotation axis has the shape of a hexagonal prism shown in Fig. 2.12f. In Fig. 2.12f, it is also shown a parallelepiped whose volume represents $1/3$ of the volume of the hexagonal prism. This parallelepiped has the shape of a primitive unit cell for the hexagonal lattice. Its basis vector $\vec{c}$ is parallel to the sixfold symmetry axis and the basis vectors $\vec{a}$ and $\vec{b}$ are lying in a plane orthogonal to this symmetry axis and are parallel to twofold rotation axes present in the hexagonal lattice. In Fig. 2.13, we show that the rhombohedron shown in Fig. 2.12e for the trigonal system and the hexagonal prism are related figures. This will be explained in more detail later. Of course in the case when $\sphericalangle(\vec{a}, \vec{b}) = \sphericalangle(\vec{c}, \vec{a}) = \sphericalangle(\vec{b}, \vec{c}) = 90°$, we are in the presence of a cube and the threefold rotation axis shown in Fig. 2.12e coincides with one of the four threefold rotation axes of the cube. To summarize, Fig. 2.12 specifies the characteristics of solid figures, which are the unit cells with the point symmetries of lattices existing in three dimensions. The point symmetry of each of the 7 figures shown in Fig. 2.12 coincides with the point symmetry of lattices belonging to one of the 7 crystal systems.

Table 2.1 Restrictions on conventional cell parameters for each crystal system. The following abbreviations are used: $\sphericalangle(\vec{b}, \vec{c}) = \alpha$, $\sphericalangle(\vec{c}, \vec{a}) = \beta$, $\sphericalangle(\vec{a}, \vec{b}) = \gamma$

Crystal system	Restrictions on conventional cell parameters a, b, c and α, β, γ
Triclinic	None
Monoclinic	$\alpha = \beta = 90°$
Orthorhombic	$\alpha = \beta = \gamma = 90°$
Tetragonal	$a = b$ $\alpha = \beta = \gamma = 90°$
Trigonal	$a = b$ $\alpha = \beta = 90°, \gamma = 120°$ ------------------------------ $a = b = c$ $\alpha = \beta = \gamma$
Hexagonal	$a = b$ $\alpha = \beta = 90°, \gamma = 120°$
Cubic	$a = b = c$ $\alpha = \beta = \gamma = 90°$

In three dimensions, the 32 crystallographic point groups and 230 space groups are classified in 7 crystal systems. Crystal structures are classified into crystal systems according to their symmetries. The symmetry of a crystal structure is determined not only by the symmetry of its conventional unit cell but also by the symmetry of the set of its basis atoms. So, on the parameters of the conventional unit cells of crystal structures belonging to a given crystal system are imposed only certain restrictions. The symmetry of the conventional unit cell has to be at least as high as the point symmetry of the lattices belonging to the crystal system to which belongs the crystal structure in consideration (in the case of the trigonal system is used a rhombohedral or a hexagonal lattice, depending on the case). The restrictions imposed on conventional cell parameters to guarantee this are summarized in Table 2.1. For instance, there are no restrictions at all on the conventional cell parameters of a crystal structure belonging to the triclinic crystal system. For this case, no constrains are necessary because symmetry axes are absent in lattices belonging to the triclinic system.

2.6 The Rhombohedron and Hexagonal Prism

A rhombohedron can be constructed inside a hexagonal prism, as shown in Fig. 2.13. We can see in the figure that two vertices of the

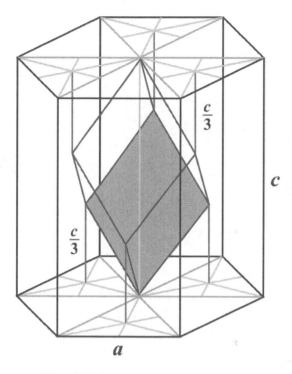

Figure 2.13 Rhombohedron constructed inside a hexagonal prism.

rhombohedron are located at the centers of the hexagonal prism bases and the other 6 form two groups with 3 vertices each. The plane defined by each group of vertices is parallel to the prism bases, which means that these vertices are at the same distance from a base. The distance between the three vertices which are closer to the top base and this base is the same as the distance between the vertices from the other group and the bottom base, and represents 1/3 of the prism height c (see Fig. 2.13). The location of the vertices belonging to each of the two groups can be determined easily as their projections on the plane of the nearer prism base (bottom or top) coincide with the geometric centers of three equilateral triangles, as shown in Fig. 2.13. We can also see in this figure that these triangles are not next to each other and the three triangles of the bottom base do not coincide with those of the top base.

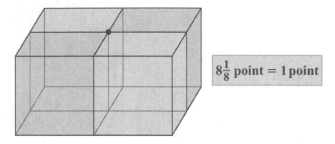

Figure 2.14 A 1/8 of the point placed in each vertex of a cubic cell belongs to this cell.

2.7 The 14 Bravais Lattices

2.7.1 *Introduction*

In this section, we will describe all the three-dimensional lattices or, more strictly speaking, lattice types. If we place lattice points at the vertices of each parallelepiped shown in Fig. 2.12, then we obtain 7 different lattices. All the points placed at the vertices of a cell contribute with 1 point to this cell. This is explained in Fig. 2.14 on the example of a cubic cell. A point placed in a vertex of a cube belongs to 8 cubes (4 of which are shown in Fig. 2.14), so 1/8 of it belongs to each cube. Since there are 8 points at the vertices of the cube, they contribute with 1 point to it, and therefore the cell is primitive.

A French physicist, Auguste Bravais (1850), demonstrated that if we place an additional point in the geometric center or additional points on the faces of the parallelepipeds from Fig. 2.12 in such a way that the set of points has the same symmetry as the parallelepiped, then we will obtain 7 new lattices or, strictly speaking, lattice types. Therefore, we have a total of 14 lattice types in three dimensions. It will be shown later that 11 of them belong to the monoclinic, orthorhombic, tetragonal, or cubic crystal systems, and to the three remaining crystal systems (triclinic, trigonal, and hexagonal) belong, in total, three lattice types. In the case of each of the 7 new lattices, the unit cell that has the same symmetry as the lattice, has more than one lattice point. Therefore, this unit cell represents a non-primitive cell of the lattice while a primitive cell of such a lattice does not have its point symmetry.

Next, we will build the 7 new lattices mentioned above, which are called the *centered Bravais lattices*.

2.7.2 The Triclinic System

In the case of the triclinic system, there is only one lattice type. Due to the absence of symmetry axes in a triclinic lattice, any primitive unit cell has the point symmetry of the lattice.

2.7.3 The Monoclinic System

In Fig. 2.15a, we have placed lattice points at the vertices of the solid figure shown in Fig. 2.12b, obtaining this way a primitive monoclinic unit cell. This cell can be additionally centered in several different ways as shown in Figs. 2.15b–d and in Fig. 2.16. In all cases, the set of lattice points has the same point symmetry as the solid figure. The cells from Figs. 2.15 and 2.16 have their unique symmetry axes parallel to the c edges. In such case, we speak of the setting with unique axis c (c-axis setting, for short). In the case of the C-face centered cell, shown in Fig. 2.15b, the centering lattice points are in the cell bases (orthogonal to the c edge). Figures 2.15c and 2.15d show also the monoclinic cells, but this time body and all-face centered (face centered, for short), respectively. There are still two more options for placing the additional lattice points within the unit cell from Fig. 2.15a. This is shown in Figs. 2.16a and 2.16b for the A-face centered and B-face centered cells, respectively. A-face centered, B-face centered, and C-face centered unit cells are known also as A-centered, B-centered, and C-centered, respectively. The symbols for the centering types of the cells shown in Figs. 2.15 and 2.16 are listed in Table 2.2.

Let us now investigate to which monoclinic lattice types belong the centered cells shown in Figs. 2.15 and 2.16. In Fig. 2.17, we demonstrate that in the lattice shown in Fig. 2.15b, we can find a primitive monoclinic unit cell, so this is, in fact, a primitive monoclinic lattice. It can be also demonstrated, and this was done in Fig. 2.18, that in the lattice from Fig. 2.15d, there is a body centered

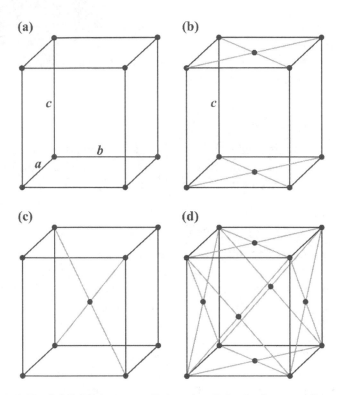

Figure 2.15 (a) Primitive monoclinic unit cell. In the figure, we have also drawn the monoclinic cells centered in three different ways: (b) C-face centered, (c) body centered, and (d) all-face centered. The c-axis setting is assumed.

monoclinic unit cell. Next, we will check the cases shown in Fig. 2.16. In each lattice type plotted in this figure, we can find a body centered monoclinic cell, like in the case of the lattice from Fig. 2.15d. This is demonstrated in Fig. 2.19 for the lattice from Fig. 2.16a. So, the lattices from Figs. 2.15c,d and 2.16a,b are mutually equivalent. The A-face centered cell will be selected as a unit cell for the centering type of the monoclinic lattice if the c-axis setting is assumed. Note that the centering type of a conventional unit cell is transferred to the lattice. The symbols of the two monoclinic lattice types are then mP (monoclinic primitive lattice) and mA (monoclinic A-centered lattice). However, the b-axis setting is considered to be the standard setting, and then the two monoclinic lattice types are mP and mC (the

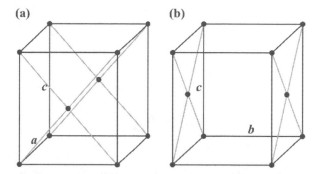

Figure 2.16   The monoclinic cells centered in two different ways: (a) *A*-face centered and (b) *B*-face centered. The *c*-axis setting is assumed.

Table 2.2   Symbols for the centering types of the cells shown in Figs. 2.15 and 2.16

Symbol	Centering type of a cell	Number of lattice points per cell
P	Primitive	1
A	*A*-face centered	2
	(*A*-centered)	
B	*B*-face centered	2
	(*B*-centered)	
C	*C*-face centered	2
	(*C*-centered)	
I	Body centered	2
	(*I*-centered)	
F	All-face centered	4
	(*F*-centered)	

C-face centered cell is selected in this case). In the case of the *mA* (*c*-axis setting) or *mC* (*b*-axis setting) lattices the smallest cell that has the point symmetry of the lattice contains 2 lattice points, while the primitive cells of these lattices do not have their point symmetry.

2.7.4 *The Orthorhombic System*

In the same way, as it was done in the previous section for the monoclinic system, we can place the lattice points within the solid figure shown in Fig. 2.12c. In each case, the resulting set of points

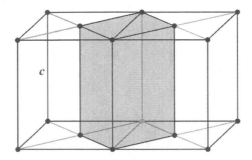

Figure 2.17 A primitive monoclinic unit cell located inside the monoclinic lattice from Fig. 2.15b. The *c*-axis setting is assumed.

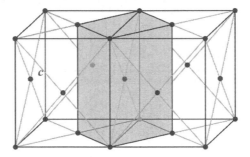

Figure 2.18 A body centered monoclinic unit cell placed inside the monoclinic lattice from Fig. 2.15d. The *c*-axis setting is assumed.

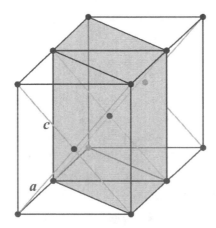

Figure 2.19 A body centered monoclinic unit cell located inside the monoclinic lattice from Fig. 2.16a. The *c*-axis setting is assumed.

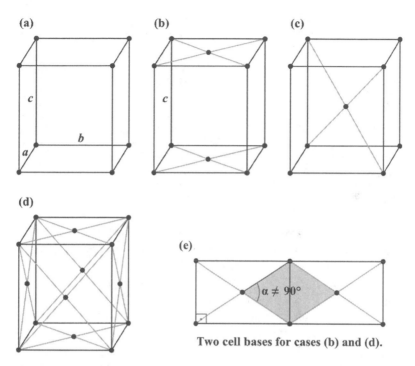

(e)

Two cell bases for cases (b) and (d).

Figure 2.20 (a) Primitive orthorhombic unit cell. In the figure, we have also drawn the orthorhombic cells centered in three different ways: (b) C-face centered, (c) body centered, and (d) all-face centered. Figure (e) shows two bases for two adjacent cells from (b) and (d). The base of a primitive (or body centered) cell of the lattice shown in (b) (or (d)) is highlighted in (e).

will have the same point symmetry as the solid figure. Since its three edges are mutually orthogonal, only the cases described in Fig. 2.20 will be considered. The rest of the cases, with A- and B-face centered cells, do not lead to any new lattice types. Contrary to what we have learned for the monoclinic system, this time, neither the case from Fig. 2.20b nor the case from Fig. 2.20d matches the cases described in Figs. 2.20a and 2.20c, respectively, since neither a primitive nor a body centered orthorhombic unit cells are present in the lattices shown in Figs. 2.20b and 2.20d, respectively (see Fig. 2.20e). Therefore, we can conclude that in the case of the orthorhombic system, there are four types of lattices: primitive (oP), C-centered (oC), body centered (oI), and face centered (oF).

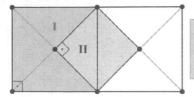

Base I corresponds to a *C*-face centered or all-face centered tetragonal unit cell, while base II corresponds to a primitive or body centered cell.

Figure 2.21 Two bases (labeled as I) of two adjacent *C*-face centered or all-face centered tetragonal unit cells. The base labeled as II corresponds in one case to a primitive tetragonal unit cell and in the other case to the body centered tetragonal cell.

2.7.5 *The Tetragonal System*

The solid figure shown in Fig. 2.12d, instead of having a rectangle at the base as was the case of the orthorhombic system, has a square. For this case, we have to analyze the same types of centering of the cell, as those shown in Figs. 2.20b–d for the orthorhombic system. Thus, the *C*-face centered, body centered, and all-face centered tetragonal cells will be considered. Here, as before, the *c* edge is orthogonal to the cell base. The presence of a fourfold rotation axis parallel to the *c* edge excludes the possibility of having *A*- and *B*-face centered cells in tetragonal lattices. It is easy to demonstrate that now the lattice represented by the *C*-face centered tetragonal unit cell is effectively a primitive tetragonal lattice and the lattice represented by the all-face centered tetragonal unit cell is just a body centered tetragonal lattice. This is shown in Fig. 2.21, where we have displayed two bases of a *C*-face centered or all-face centered tetragonal unit cells. One of those bases is labeled as I, whereas the square marked as II is the base of a primitive or body centered tetragonal cell.

To conclude, we can say that in the case of the tetragonal system, there are two types of lattices: *tP* and *tI*.

2.7.6 *The Cubic System*

We now move to the case of a cubic system. The search for the possible lattices belonging to this system will be held using again Fig. 2.20. This time, the only relevant cases are those described in Figs. 2.20a,c,d, since, due to the point symmetry, those are the only

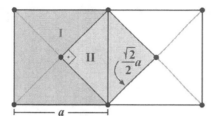

A body centered cell in the face centered cubic lattice does not have all the edges of the same longitude.

Figure 2.22 Two bases (labeled as I) of two adjacent all-face centered cubic unit cells. The base labeled as II corresponds to a non-cubic body centered unit cell.

cases that can be *a priori* expected in the lattices of the cubic system (remember that in the present consideration, all the cells shown in Fig. 2.20 are cubes). In the cubic lattice with the same arrangement of lattice points as shown in Fig. 2.20d, it is excluded the presence of a cubic body centered unit cell, as explained in Fig. 2.22.

Thus, we can conclude that in the case of the cubic system, there are 3 types of lattices: *cP*, *cI*, and *cF*. The primitive cubic lattice is also called simple cubic (*sc*) and the body and face centered cubic lattices are commonly abbreviated as *bcc* (body centered cubic) and *fcc* (face centered cubic), respectively.

2.7.7 The Trigonal and Hexagonal Systems

There is only one type of lattice, *hP*, in the hexagonal crystal system, while in the case of the trigonal system there are two types of lattices: *hP* and *hR*. Later, we will explain the origin of the symbol *hR* used for the rhombohedral lattice.

Finally, we may say that there are all together 14 types of lattices in three dimensions, called the Bravais lattices. The 14 Bravais lattices are collected in Fig. 2.23.

2.7.8 Symbols for the Bravais Lattices

In Table 2.3 are summarized the symbols for the 14 Bravais lattices. We can observe in this table, that the Bravais lattices are classified in 6 crystal families, that are symbolized by lower case letters *a*, *m*, *o*, *t*, *h*, and *c* (see column two of Table 2.3). The second classification is according to the discussed by us 7 crystal systems. We can see

Table 2.3 Symbols for the 14 Bravais lattices

Crystal family	Symbol	Crystal system	Bravais lattice symbol
Triclinic (anorthic)	a	Triclinic	aP
Monoclinic	m	Monoclinic	mP
			mS (mA, mB, mC)
Orthorhombic	o	Orthorhombic	oP
			oS (oA, oB, oC)
			oI
			oF
Tetragonal	t	Tetragonal	tP
			tI
Hexagonal	h	Trigonal	hP, hR
		Hexagonal	hP
Cubic	c	Cubic	cP
			cI
			cF

in the table that the classification according to crystal families and crystal systems is the same except for the hexagonal family, which collects two crystal systems: trigonal and hexagonal. The two parts of the Bravais lattice symbol are: first, the symbol of the crystal family and second, a capital letter (P, S, I, F, R) designating the Bravais lattice centering. As a reminder, the symbol P is given to primitive lattices. The symbol S denotes a one-face centered lattice (mS and oS are the standard, setting independent, symbols for the one-face centered monoclinic and orthorhombic Bravais lattices, respectively). Instead of S also the A, B, or C symbols are used, describing lattices centered at the corresponding A, B, or C faces. Symbols F and I are designated for face centered and body centered Bravais lattices, respectively. Finally, the symbol R is used for the rhombohedral lattice.

2.7.9 Final Remarks

To conclude, we can say that the three-dimensional lattices are classified in 7 crystal systems. The lattices belonging to a given crystal system have the same point symmetry (with the exception of the hP and hR lattices in the case of the trigonal system). On the other hand the lattices that have the same point symmetry are still classified due to their space symmetry. The lattices that have

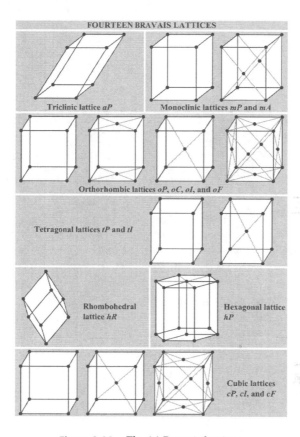

Figure 2.23 The 14 Bravais lattices.

the same space symmetry have the common name that corresponds to one of the Bravais lattices. We could see that half of the Bravais lattices appear as centered ones. This means that the smallest unit cells that have the same point symmetry as the lattices contain more than one lattice point. However, it is important to point out that for each of the 14 Bravais lattices it is possible to choose a unit cell that contains only one lattice point, that is, a primitive unit cell. The basis vectors $\vec{a}_1$, $\vec{a}_2$, $\vec{a}_3$ that define such a cell are primitive translation vectors of the Bravais lattice. Finally, a Bravais lattice represents a set of points whose positions are given by vectors $\vec{R}$ defined as

$$\vec{R} = n_1\vec{a}_1 + n_2\vec{a}_2 + n_3\vec{a}_3, \qquad (2.1)$$

where n_1, n_2, $n_3 \in \mathbb{Z}$. The 14 Bravais lattices are shown in Fig. 2.23.

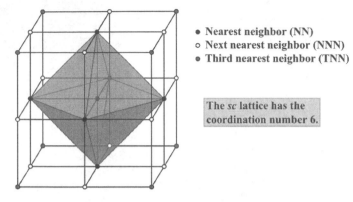

- Nearest neighbor (NN)
- Next nearest neighbor (NNN)
- Third nearest neighbor (TNN)

> The *sc* lattice has the coordination number 6.

Figure 2.24 The NNs, NNNs, and TNNs of a lattice point in a *sc* lattice. The 6 NNs of a lattice point placed in the center of the large cube are at the vertices of the regular octahedron. The 12 NNNs are in the middle of the large cube edges and the 8 TNNs are in its vertices.

2.8 Coordination Number

Since all the lattice points in a lattice have equivalent positions in space, they have identical surroundings. Therefore, each lattice point has the same number of NNs (points that are the closest to it) and this number, called the *coordination number*, is a characteristic of a given lattice. Moreover, in the case of cubic lattices the coordination number is a characteristic of a given Bravais lattice. However, in the literature, we find more often an alternative definition in which the coordination number is the number of the NNs of an atom in a crystal. Of course, in the case of one-atom basis, the two definitions are equivalent.

On occasions, the information about the next nearest neighbors (NNNs) and even the third nearest neighbors (TNNs) of a lattice point is also important. Figure 2.24 shows the NNs, NNNs, and TNNs of a lattice point in the *sc* lattice. In this figure, the NNs of a lattice point placed in the center of the large cube (built of 8 smaller cubes) are placed at the vertices of a regular octahedron. Since the octahedron has 6 vertices, the coordination number for the *sc* lattice is 6. The NNNs are in the middle of the 12 edges of the large cube, so there are 12 NNNs of a lattice point in the *sc* lattice. The TNNs of the lattice point in consideration are at the vertices of the large cube, so the number of them is 8. The NN distance in the *sc* lattice is equal to the lattice parameter a, the NNN distance is $\sqrt{2}a$, and the TNN distance is $\sqrt{3}a$.

2.9 Body Centered Cubic Lattice

Figure 2.25 shows three examples of a set of primitive translation vectors that define the primitive unit cell of the *bcc* lattice. In these three cases at least one of the vectors involves two "types" of lattice points, namely, those from cube vertices and those from cube centers. This, of course, is essential in the case of a primitive cell, since such a cell with its only one lattice point has to reproduce the entire lattice. The primitive cell defined by vectors $\vec{a}_1$, $\vec{a}_2$, and $\vec{a}_3$ in Fig. 2.25c is shown in Fig. 2.26. In this figure, it is drawn a rhombohedron which represents the most symmetric primitive unit cell of the *bcc* lattice. We can also see in this figure that one diagonal of the rhombohedron is lying along one of the diagonals of the cube. Those diagonals represent a threefold rotation axis of each cell. This is the unique threefold axis of the rhombohedron, while the cube has still three more such axes. The angles between the basis vectors $\vec{a}_1$, $\vec{a}_2$, and $\vec{a}_3$ shown in Figs. 2.25c and 2.26 are the same: $\sphericalangle(\vec{a}_1, \vec{a}_2) = \sphericalangle(\vec{a}_3, \vec{a}_1) = \sphericalangle(\vec{a}_2, \vec{a}_3) = 109°28'$.

Next, we will calculate the volume Ω_0 of the primitive unit cell and compare it with the volume of the cube. The volume of this cell is given by

$$\Omega_0 = (\vec{a}_1 \times \vec{a}_2) \cdot \vec{a}_3 = \begin{vmatrix} \hat{x} & \hat{y} & \hat{z} \\ \frac{1}{2}a & -\frac{1}{2}a & \frac{1}{2}a \\ \frac{1}{2}a & \frac{1}{2}a & -\frac{1}{2}a \end{vmatrix} \cdot \left(-\frac{1}{2}a\hat{x} + \frac{1}{2}a\hat{y} + \frac{1}{2}a\hat{z} \right)$$

$$= \left(0\hat{x} + \frac{1}{2}a^2\hat{y} + \frac{1}{2}a^2\hat{z} \right) \cdot \left(-\frac{1}{2}a\hat{x} + \frac{1}{2}a\hat{y} + \frac{1}{2}a\hat{z} \right)$$

$$= \frac{1}{4}a^3 + \frac{1}{4}a^3 = \frac{1}{2}a^3, \tag{2.2}$$

and the volumes ratio is

$$\frac{V_{\text{cube}}}{\Omega_0} = \frac{a^3}{\frac{1}{2}a^3} = 2. \tag{2.3}$$

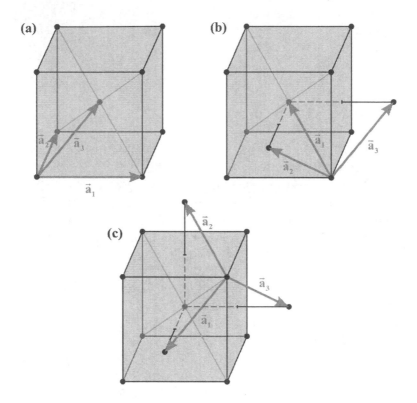

Figure 2.25 Three sets of primitive translation vectors of the *bcc* lattice.

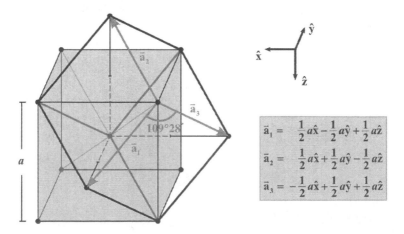

$$\vec{a}_1 = \frac{1}{2}a\hat{x} - \frac{1}{2}a\hat{y} + \frac{1}{2}a\hat{z}$$

$$\vec{a}_2 = \frac{1}{2}a\hat{x} + \frac{1}{2}a\hat{y} - \frac{1}{2}a\hat{z}$$

$$\vec{a}_3 = -\frac{1}{2}a\hat{x} + \frac{1}{2}a\hat{y} + \frac{1}{2}a\hat{z}$$

Figure 2.26 A primitive rhombohedral unit cell of the *bcc* lattice.

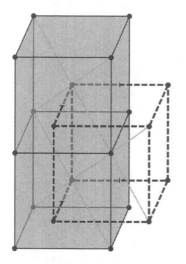

The points that are in the centers of the gray colored cubes coincide with the vertices of cubes that have dashed edges, and *vice versa*: the points that are in the centers of the "dashed" cubes overlap with the vertices of the gray colored ones.

Figure 2.27 Demonstration of the equivalence of the two lattice points within the cubic unit cell of the *bcc* lattice.

A primitive unit cell of the *bcc* lattice has one lattice point while the cubic cell has two points. The ratio, given by Eq. 2.3, between the cell volumes is equal to the ratio between the numbers of points belonging to them. Therefore, the same volume Ω_0 corresponds to each lattice point.

Let us now refer to the fact that the two points that are within the cubic *I* unit cell have equivalent positions in the *bcc* lattice. This is shown and explained in Fig. 2.27. In Fig. 2.28, we show the NNs of a point of the *bcc* lattice. This lattice has a coordination number 8.

We will now consider the lattice points within the cubic cell of the *bcc* lattice. The point being a sum of eight fractions (see Fig. 2.29a) is associated with the cell origin, as shown in Fig. 2.29b. The coordinate triplet of the second lattice point within the cubic cell is $\frac{1}{2}, \frac{1}{2}, \frac{1}{2}$, where the coordinates are expressed in units of a. Vector $\vec{t} = 1/2\vec{a} + 1/2\vec{b} + 1/2\vec{c}$ represents one of the shortest translation vectors of the *bcc* lattice. This vector is defined by the two lattice points belonging to the cubic unit cell (see Fig. 2.29b). At least one such vector has to appear in each set of basis vectors that define a primitive unit cell of the *bcc* lattice (see Fig. 2.25), since both, the lattice point located at the center of the cubic cell and the point from

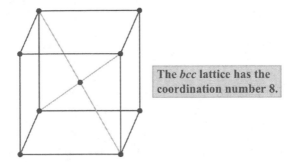

Figure 2.28 The lattice points from the vertices of the cubic unit cell of the *bcc* lattice represent the NNs of the lattice point that is in the center of the cell.

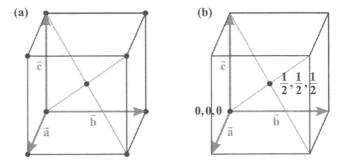

Figure 2.29 (a) Conventional unit cell of the *bcc* lattice. Each of the lattice points located at the vertices contributes with 1/8 to the unit cell so the cell contains 2 lattice points. (b) The coordinate triplets of the two lattice points within the cubic cell. The point, which is a sum of eight fractions, is placed at the origin of the cell. The coordinates are expressed in units of *a*.

its vertices, are then represented by the lattice point from vertices of the primitive unit cell.

2.10 Face Centered Cubic Lattice

First, let us consider the lattice points within the cubic unit cell of the *fcc* lattice. The two lattice points placed in the *A*-faces (orthogonal to the basis vector $\vec{a}$ in Fig. 2.30a) contribute to the cubic cell with half of the point each. We will represent these two fractions with one

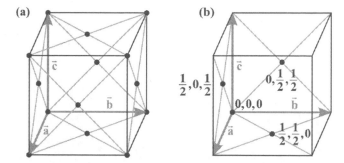

Figure 2.30 (a) Face centered cubic unit cell of the *fcc* lattice. (b) Coordinate triplets of the four lattice points within the cubic cell. The coordinates are expressed in units of *a*, which is the length of the cube edge.

lattice point placed in the A face that contains the origin of the cell (see Fig. 2.30b), that is, at the shortest distance from the origin. The coordinate triplets of this point is $0, \frac{1}{2}, \frac{1}{2}$, where the coordinates are expressed in units of a, which is the length of the basis vectors. In the similar way, we can place the points in the B- and C-faces (see Fig. 2.30b). Finally, the coordinate triplets of the four lattice points belonging to the cubic cell of the *fcc* lattice are: $0, 0, 0$; $0, \frac{1}{2}, \frac{1}{2}$; $\frac{1}{2}, 0, \frac{1}{2}$; and $\frac{1}{2}, \frac{1}{2}, 0$.

Figure 2.31 shows the most symmetric primitive unit cell of the *fcc* lattice. It is defined by the basis vectors $\vec{a}_1$, $\vec{a}_2$, and $\vec{a}_3$. Each of them represents one of the shortest translation vectors of the *fcc* lattice. In the definition of the vectors $\vec{a}_1$, $\vec{a}_2$, and $\vec{a}_3$ are involved all the four lattice points belonging to the cubic unit cell and this guarantees that the primitive cell can reproduce the entire lattice. The primitive unit cell shown in Fig. 2.31 takes on the shape of a rhombohedron that is inscribed in the cubic cell. The threefold rotation axis of the rhombohedron coincides with one of the threefold rotation axis of the cube. The lattice points that define this axis are at the vertices of the two cells, while the rest of the rhombohedron vertices coincide with the centers of the cube faces.

For the cell form Fig. 2.31 it is easy to show that

$$\sphericalangle(\vec{a}_1, \vec{a}_2) = \sphericalangle(\vec{a}_3, \vec{a}_1) = \sphericalangle(\vec{a}_2, \vec{a}_3) = 60°. \qquad (2.4)$$

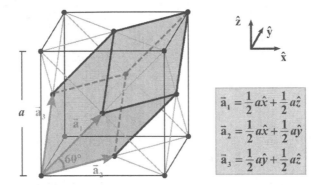

Figure 2.31 A primitive rhombohedral unit cell of the *fcc* lattice.

$$\vec{a}_1 = \frac{1}{2}a\hat{x} + \frac{1}{2}a\hat{z}$$

$$\vec{a}_2 = \frac{1}{2}a\hat{x} + \frac{1}{2}a\hat{y}$$

$$\vec{a}_3 = \frac{1}{2}a\hat{y} + \frac{1}{2}a\hat{z}$$

Indeed, since $a_1 = a_2 = a_3 = (\sqrt{2}/2)a$ we have

$$\vec{a}_1 \cdot \vec{a}_2 = a_1 a_2 \cos \sphericalangle(\vec{a}_1, \vec{a}_2) = \frac{2}{4}a^2 \cos \sphericalangle(\vec{a}_1, \vec{a}_2) \qquad (2.5)$$

and using the vector components we have also

$$\vec{a}_1 \cdot \vec{a}_2 = a_{1x}a_{2x} + a_{1y}a_{2y} + a_{1z}a_{2z} = \frac{1}{4}a^2, \qquad (2.6)$$

and then comparing the two expressions for the scalar product $\vec{a}_1 \cdot \vec{a}_2$, we obtain

$$\frac{2}{4}a^2 \cos \sphericalangle(\vec{a}_1, \vec{a}_2) = \frac{1}{4}a^2 \Rightarrow \cos \sphericalangle(\vec{a}_1, \vec{a}_2) = \frac{1}{2}. \qquad (2.7)$$

Repeating the same procedure as done in Eqs. 2.5–2.7, for all the vector pairs, we finally get Eq. 2.4. So the basis vectors $\vec{a}_1$, $\vec{a}_2$, and $\vec{a}_3$ of a primitive rhombohedral unit cell of the *fcc* lattice are at angles of 60° to each other.

Let us now calculate the volume, Ω_0, of the primitive unit cell of the *fcc* lattice and compare it with the volume of the cubic cell. We have

$$\Omega_0 = (\vec{a}_1 \times \vec{a}_2) \cdot \vec{a}_3 = \begin{vmatrix} \hat{x} & \hat{y} & \hat{z} \\ \frac{1}{2}a & 0 & \frac{1}{2}a \\ \frac{1}{2}a & \frac{1}{2}a & 0 \end{vmatrix} \cdot \left(\frac{1}{2}a\hat{y} + \frac{1}{2}a\hat{z}\right)$$

$$= \left(-\frac{1}{4}a^2\hat{x} + \frac{1}{4}a^2\hat{y} + \frac{1}{4}a^2\hat{z}\right) \cdot \left(\frac{1}{2}a\hat{y} + \frac{1}{2}a\hat{z}\right)$$

$$= \frac{1}{8}a^3 + \frac{1}{8}a^3 = \frac{1}{4}a^3. \qquad (2.8)$$

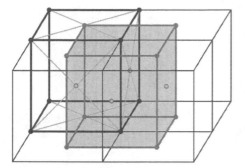

In the figure, we have two types of cubes: gray colored and uncolored. The "black" lattice points, ● , that are at the vertices of the uncolored cubes are in the centers of the bases of the gray colored cubes, and *vice versa*. The "light gray" lattice points, ○ , are in the centers of the faces of both types of cubes.

Figure 2.32 Demonstration of the equivalence of all lattice points in the cubic cell of the *fcc* lattice.

The primitive unit cell has one lattice point while the cubic cell contains four lattice points, so the ratio between the volumes of these cells

$$\frac{V_{\text{cube}}}{\Omega_0} = \frac{a^3}{\frac{1}{4}a^3} = 4 \tag{2.9}$$

is equal to the ratio between the numbers of lattice points in them.

Next, we will demonstrate that different lattice points within the cubic unit cell of the *fcc* lattice have equivalent positions in this lattice. This is shown in Fig. 2.32. In the explanation, we are using two sets of cubes. The second set of cubes (represented by a gray colored cube in Fig. 2.32) is obtained by translating the first one in the direction of a diagonal from its bases by half of the diagonal length. The correspondence of the lattice points in the two sets of cubes is explained in Fig. 2.32. Upon making a similar translation, but now in a plane which coincides with the side faces of one of the two sets of cubes, the result will be that the points that are in the middle of the faces of the two types of cubes (from the two sets) will occupy the positions of the points at the vertices. In this manner, we can demonstrate the equivalence between the positions of lattice points at the vertices and faces of the cube.

Since all the lattice points in the *fcc* lattice have equivalent positions, the neighborhood of each lattice point is the same and therefore, each point has the same number of NNs. We will consider the neighborhood of a lattice point from the face of a cube. This is shown in Fig. 2.33 for the central lattice point of the draw. As

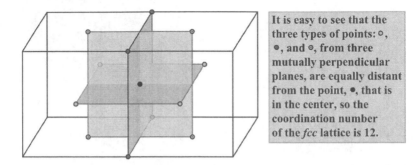

It is easy to see that the three types of points: ○, ●, and ⊙, from three mutually perpendicular planes, are equally distant from the point, ●, that is in the center, so the coordination number of the *fcc* lattice is 12.

Figure 2.33 Nearest neighbors of a lattice point in the *fcc* lattice.

explained in this figure the coordination number of the *fcc* lattice is 12.

2.11 Rhombohedral Unit Cell in a Cubic Lattice

2.11.1 *Introduction*

We have already learned in Sections 2.9 and 2.10 that the most symmetric primitive unit cells of both the *bcc* and the *fcc* lattices have shapes of rhombohedrons. However, a rhombohedron with one lattice point represents, at first, the conventional unit cell of the rhombohedral Bravais lattice, and now we know that when the basis vectors $\vec{a}_1$, $\vec{a}_2$, and $\vec{a}_3$ of a primitive rhombohedral unit cell are at angles of 60° or 109°28' to each other, then this cell is a primitive unit cell of a lattice belonging to the cubic system, which means that such lattice possesses a higher point symmetry than the symmetry of a rhombohedral lattice. Moreover, the presence of a rhombohedral unit cell with its threefold symmetry axis in a cubic lattice is not surprising, since this lattice possesses threefold symmetry axes. Next we will show a centered rhombohedral unit cell in the *sc* lattice.

2.11.2 *A Rhombohedral Unit Cell of the sc Lattice*

Besides of the primitive rhombohedral unit cells there are, of course, centered rhombohedral unit cells in lattices belonging to the cubic system. Figure 2.34 shows such a unit cell in the *sc* lattice. This

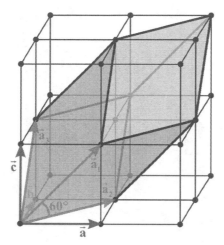

Figure 2.34 A body centered rhombohedral unit cell of the *sc* lattice.

cell contains two lattice points. One of them is placed at the origin of the cell and the position of the other is given by the vector $(1/2)(\vec{a}_1 + \vec{a}_2 + \vec{a}_3)$, where $\vec{a}_1$, $\vec{a}_2$, and $\vec{a}_3$ are the vectors defining the rhombohedral unit cell in Fig. 2.34.

Table 2.4 resumes the information about the two types of unit cells shown in Fig. 2.34, that is, the cubic P defined by basis vectors $\vec{a}$, $\vec{b}$, and $\vec{c}$ and the rhombohedral I defined by vectors $\vec{a}_1$, $\vec{a}_2$, and $\vec{a}_3$. The volume of the body centered rhombohedral cell is two times the volume of the primitive cubic cell (see the number of lattice points within each cell). The fact that in the *sc* lattice is present a rhombohedral unit cell defined by vectors which are at angles of 60° to each other suggests that a *sc* crystal structure may be described using the *fcc* lattice, what will be considered below.

Table 2.4 The characteristics of the two unit cells, shown in Fig. 2.34, of the *sc* lattice

Simple Cubic Lattice		
Unit cell type	Cell parameters	Number of lattice points per cell
Cubic P	a_c	1
Rhombohedral I	$a_r = \sqrt{2}a_c,\quad \alpha_r = 60°$	2

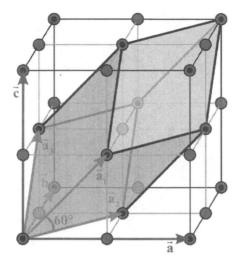

Figure 2.35 Simple cubic crystal structure. Face centered cubic lattice is used to describe this structure. The cubic F and the rhombohedral P unit cells of this lattice, defined by the basis vectors $\vec{a}$, $\vec{b}$, $\vec{c}$ and $\vec{a}_1$, $\vec{a}_2$, $\vec{a}_3$, respectively, are shown. The cubic F unit cell contains 8 atoms, while the rhombohedral P unit cell contains 2 atoms.

2.11.3 *Simple Cubic Crystal Structure*

Let us now consider a *sc* crystal structure. The conventional unit cell of this structure is cubic P with the origin at the center of an atom. It contains 1 atom, and therefore

$$sc \text{ crystal structure} \equiv sc \text{ lattice} + 1\text{-atom basis.}$$

However, the F-centered cubic unit cell may be also used to describe the *sc* structure, since

$$sc \text{ crystal structure} \equiv fcc \text{ lattice} + 2\text{-atom basis.}$$

Indeed, in the *sc* crystal structure shown in Fig. 2.35, the cubic F unit cell defined by basis vectors $\vec{a}$, $\vec{b}$, and $\vec{c}$ has 8 atoms and 4 lattice points, it means, two atoms per lattice point. Figure 2.36a shows additionally the rhombohedral P unit cell defined in Fig. 2.35, with two atoms in it. The center of one of the basis atoms coincides with the origin **O** of the cell, while Fig. 2.36b shows an alternative choice, **O′**, for the rhombohedral P cell origin. The *fcc* lattice will be used in Section 3.14 to describe the high pressure (25 GPa and room

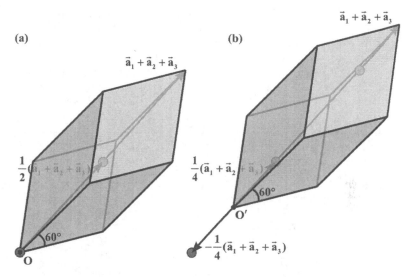

Figure 2.36 The *fcc* lattice used to describe the *sc* structure. The figure shows a primitive rhombohedral unit cell of the *fcc* lattice (defined by vectors $\vec{a}_1, \vec{a}_2, \vec{a}_3$ in Fig. 2.35) with two atoms in it. (a) The cell origin overlaps the center of an atom. (b) An alternative choice of the cell origin.

temperature) phase of arsenic whose structure is of *sc* type. The use of the *fcc* lattice makes easier the comparison of this structure with the arsenic structure at normal conditions.

2.11.4 *Interpretation of Data for As, Sb, Bi, and Hg*

We will use now the considerations made in the previous section to show that the structures of arsenic, antimony, and bismuth (in their α phases) are at normal conditions close to the *sc* structure. In order to do this, we have to realize first that a rhombohedron may be obtained by stretching a cube along one of its body diagonals. In presence of such distortion, the *sc* crystal structure transforms to the crystal structure with rhombohedral space-group symmetry. Then, the conventional cubic *P* unit cell of the *sc* structure becomes a rhombohedral *P* unit cell with angles between its basis vectors smaller than 90°, and the rhombohedron defined by vectors $\vec{a}_1, \vec{a}_2$, and $\vec{a}_3$ in Fig. 2.35 changes into a different rhombohedron with angles between its basis vectors smaller than 60°. The experimental

Table 2.5 Experimental lattice parameters obtained under normal conditions for arsenic, antimony, and bismuth. It is also given the value for mercury at normal pressure and temperature 227 K. The conventional basis vectors $\vec{a}_1$, $\vec{a}_2$, and $\vec{a}_3$ define a rhombohedral P unit cell that in the case of As, Sb, and Bi contains two atoms. The data about the location of basis atoms are given with respect to the conventional cell origin, whose location is similar to that from Fig. 2.36b. In the case of Hg the basis is composed of one atom

Element	Lattice parameters a_r(Å), α_r	Number of atoms in a rhombohedral unit cell	Coordinates of the basis atoms given in terms of vector $(\vec{a}_1 + \vec{a}_2 + \vec{a}_3)$
α-As	$a_r = 4.131$ $\alpha_r = 54°10'$	2	$x = \pm 0.226$
α-Sb	$a_r = 4.50661$ $\alpha_r = 57°6'$	2	$x = \pm 0.233$
α-Bi	$a_r = 4.7459$ $\alpha_r = 57°14'$	2	$x = \pm 0.237$
α-Hg (227 K)	$a_r = 3.005$ $\alpha_r = 70°19'$	1	$x = 0$

data for α-As, α-Sb, and α-Bi provided in Table 2.5 can be then interpreted as follows. The three elements crystallize in the so-called α-As crystal structure. In each case the conventional rhombohedral unit cell contains 2 atoms placed with respect to the origin, as was proposed in Fig. 2.36b for the sc structure. The angles between the basis vectors $\vec{a}_1$, $\vec{a}_2$, and $\vec{a}_3$ are close to 60° and the position vectors of the basis atoms, $x(\vec{a}_1 + \vec{a}_2 + \vec{a}_3)$, with respect to the origin (see Fig. 2.36b) are nearly $(\pm 1/4)$ $(\vec{a}_1 + \vec{a}_2 + \vec{a}_3)$, which is the case of the sc crystal structure. Therefore, the crystal structure of each of the three elements is close to the sc structure.

To compare, Table 2.5 reports also the data for the crystal structure of mercury at normal pressure and temperature 227 K (conditions, in which mercury has a structure with rhombohedral space-group symmetry). Its conventional unit cell is rhombohedral P, and contains only one atom. The cell parameter α_r is such that the α-Hg crystal structure is not close to any cubic structure, however, it could be considered as derived from the fcc structure by applying a strong compression along one of its threefold axes.

2.12 Rhombohedral Lattice

A conventional cell of the rhombohedral lattice (that belongs to the trigonal system), has the shape of a rhombohedron. We know already that a rhombohedron can be constructed inside a hexagonal prism (see Fig. 2.13). In such a construction, the sixfold rotation axis of the hexagonal prism becomes a threefold rotation axis of the rhombohedron. It is easy to realize that a rhombohedral lattice could be considered as a centered hexagonal one. The presence of additional lattice points in the rhombohedral lattice, with respect to the hexagonal lattice, reduces the sixfold rotation axis of the hexagonal prism to a threefold one, as shown in Fig. 2.37. We can see in this figure that the rhombohedral lattice points that are inside the hexagonal prism define two equilateral triangles in planes orthogonal to the sixfold hexagonal prism symmetry axis. The axis is crossing these planes at the geometric centers of the triangles. Such an arrangement of the rhombohedral lattice points, which are inside the hexagonal prism, reduces a sixfold rotation axis of the hexagonal lattice to a threefold rotation axis of the rhombohedral lattice. The basis vectors $\vec{a}_h$, $\vec{b}_h$, and $\vec{c}_h$ in Fig. 2.37 define a rhombohedrally

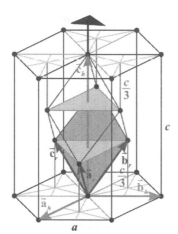

The obverse setting of a triple hexagonal cell in relation to a primitive rhombohedral cell;
$$\vec{a}_h = \vec{a}_r - \vec{b}_r, \ \vec{b}_h = \vec{b}_r - \vec{c}_r,$$
$$\vec{c}_h = \vec{a}_r + \vec{b}_r + \vec{c}_r.$$

Figure 2.37 Primitive rhombohedral and a R-centered hexagonal unit cells of a rhombohedral lattice. The centering points, within the hexagonal cell, reduce the sixfold rotation axis of the hexagonal prism to a threefold rotation axis.

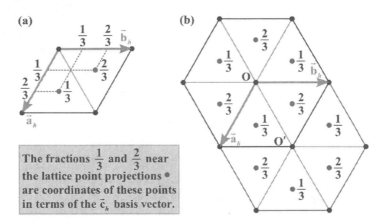

(a)

(b)

The fractions $\frac{1}{3}$ and $\frac{2}{3}$ near the lattice point projections • are coordinates of these points in terms of the $\vec{c}_h$ basis vector.

Figure 2.38 (a) Projections of the centering points of the triple hexagonal cell from Fig. 2.37 on the base of the cell. The coordinates of these points are given in terms of the basis vectors $\vec{a}_h$, $\vec{b}_h$, and $\vec{c}_h$. (b) Projections of the 6 points that are inside the hexagonal prism on its base. The fractions $\frac{1}{3}$ and $\frac{2}{3}$ near the lattice point projections are coordinates of these points in terms of the basis vector $\vec{c}_h$. The hexagonal prism base translated by a translation vector $(\vec{a}_h + \vec{b}_h)$ is also shown.

centered hexagonal unit cell (*R*-centered hexagonal cell) for a rhombohedral lattice, and the symbol of this lattice is *hR*. The *R*-centered hexagonal unit cell is called a *triple hexagonal cell*. It contains three lattice points and represents another option for a conventional unit cell of the rhombohedral lattice.

In Fig. 2.38a, we show the projections of the centering points of the triple hexagonal cell from Fig. 2.37 on the cell base. Whereas Fig. 2.38b shows the projections on the prism base of 6 rhombohedral lattice points that are inside the hexagonal prism from Fig. 2.37. In this figure **O** represents the origin of the triple hexagonal unit cell defined by basis vectors $\vec{a}_h$, $\vec{b}_h$, and $\vec{c}_h$. The coordinate of each point, in terms of the basis vector $\vec{c}_h$, is shown next to the lattice point projection and is expressed in units of *c*. In Fig. 2.38b, there is also shown the base of the hexagonal prism in consideration translated by a translation vector $(\vec{a}_h + \vec{b}_h)$. For this case, the origin of the triple hexagonal unit cell changes from **O** to **O′**. We can see in Fig. 2.38b that by translating a rhombohedral

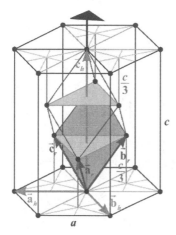

The reverse setting of a triple hexagonal cell in relation to a primitive rhombohedral cell;
$$\vec{a}_h = \vec{c}_r - \vec{b}_r, \; \vec{b}_h = \vec{a}_r - \vec{c}_r,$$
$$\vec{c}_h = \vec{a}_r + \vec{b}_r + \vec{c}_r.$$

Figure 2.39 The reverse setting of a triple hexagonal cell in relation to the primitive rhombohedral cell.

lattice to a translation vector $(\vec{a}_h + \vec{b}_h)$ we obviously obtain the same rhombohedral lattice.

The setting of the triple hexagonal unit cell in relation to the primitive rhombohedral unit cell is not unique. In Fig. 2.37 is shown the *obverse setting* of the triple hexagonal unit cell, and in Fig. 2.38a, as we know, is displayed the projection of this cell onto the base of the cell. If we propose the basis vectors $\vec{a}_h$, $\vec{b}_h$, and $\vec{c}_h$ for the hexagonal cell in the way done in Fig. 2.39, then we obtain the *reverse setting* of the triple hexagonal unit cell. The location of the centering points in the hexagonal cell depends on the setting in consideration.

Figure 2.40 shows two triple hexagonal *R* cells. The cell from Fig. 2.40a is in obverse setting, while the cell from Fig. 2.40b is in reverse setting. Both figures show the location of the three lattice points within the hexagonal *R* unit cell. We can observe that the coordinates of the common centering point with respect to the basis vectors $\vec{a}_h$ and $\vec{b}_h$ are in each case different (of course, the vectors $\vec{a}_h$ and $\vec{b}_h$ are also different).

Summarizing, we have considered here two types of conventional unit cells—rhombohedral *P* and triple hexagonal *R*—for a rhombohedral lattice. It is probably more convenient to describe a rhombohedral lattice using a triple hexagonal *R* cell, since the

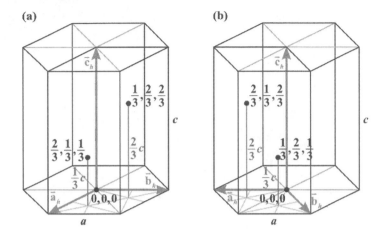

(a) **(b)**

Figure 2.40 Coordinate triplets of the three points within the triple hexagonal R unit cell in obverse setting (a) and in reverse setting (b). The coordinates are expressed in units of a and c.

hexagonal axes are easier to visualize. The relations between the basis vectors that define a rhombohedral P cell and the ones that define a triple hexagonal R cells are $\vec{a}_h = \vec{a}_r - \vec{b}_r$, $\vec{b}_h = \vec{b}_r - \vec{c}_r$, $\vec{c}_h = \vec{a}_r + \vec{b}_r + \vec{c}_r$, in the case of the obverse setting and $\vec{a}_h = \vec{c}_r - \vec{b}_r$, $\vec{b}_h = \vec{a}_r - \vec{c}_r$, $\vec{c}_h = \vec{a}_r + \vec{b}_r + \vec{c}_r$, in the case of the reverse setting.

2.13 Triple Hexagonal R Cell in the Cubic Lattice

The centered cubic lattices (*bcc* and *fcc*) possess primitive rhombohedral cells so it is natural to introduce for them the triple hexagonal cells. Therefore, the *bcc* and *fcc* lattices may be described in terms of cubic, rhombohedral, and hexagonal axes by using cubic (body centered or face centered), primitive rhombohedral, and triple hexagonal R unit cells, respectively. In Fig. 2.41 we show the three types of cells of the *bcc* lattice by putting them all together and with a common origin. Some information about those cells is listed in Table 2.6 and in Table 2.7 is listed the same information, but for the three types of unit cells of the *fcc* lattice.

CUBIC *I*, RHOMBOHEDRAL *P*, AND TRIPLE HEXAGONAL *R* UNIT CELLS FOR THE BCC LATTICE

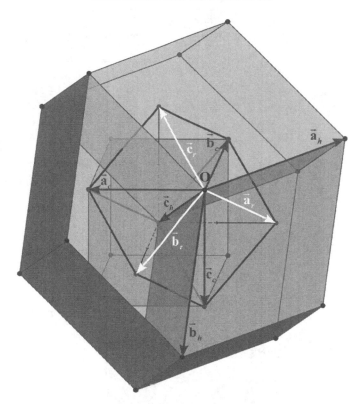

The vectors:

$$\vec{a}_h = \vec{a}_r - \vec{b}_r,$$
$$\vec{b}_h = \vec{b}_r - \vec{c}_r, \text{ and}$$
$$\vec{c}_h = \vec{a}_r + \vec{b}_r + \vec{c}_r$$

define the triple hexagonal unit cell in the obverse setting in relation to the primitive rhombohedral unit cell defined by vectors $\vec{a}_r, \vec{b}_r, \vec{c}_r$.

Figure 2.41   Three types of unit cells of the *bcc* lattice. Each of the three triple hexagonal *R* cells shown in the figure is defined by basis vectors $\vec{a}_h$, $\vec{b}_h$, and $\vec{c}_h$ or their linear combinations. Inside the hexagonal prism there is a rhombohedral *P* unit cell defined by basis vectors $\vec{a}_r$, $\vec{b}_r$, and $\vec{c}_r$. Besides that, there is a cubic *I* cell of the *bcc* lattice defined by vectors $\vec{a}_c$, $\vec{b}_c$, $\vec{c}_c$. All three unit cells have the same origin **O**.

Table 2.6 Basic information about three types of unit cells of the *bcc* lattice

Body Centered Cubic Lattice		
Unit cell type	**Cell parameters**	**Number of lattice points per cell**
Cubic *I*	a_c	2
Rhombohedral *P*	$a_r = (\sqrt{3}/2)a_c$ $\alpha_r = 109°28'$	1
Triple hexagonal *R*	$a_h = \sqrt{2}a_c$ $c_h = (\sqrt{3}/2)a_c$	3

Table 2.7 Basic information about three types of unit cells of the *fcc* lattice

Face Centered Cubic Lattice		
Unit cell type	**Cell parameters**	**Number of lattice points per cell**
Cubic *F*	a_c	4
Rhombohedral *P*	$a_r = (\sqrt{2}/2)a_c$ $\alpha_r = 60°$	1
Triple hexagonal *R*	$a_h = (\sqrt{2}/2)a_c$ $c_h = \sqrt{3}a_c$	3

2.14 Wigner–Seitz Cell

2.14.1 *Introduction*

The primitive unit cells that we have considered, until now, for the case of centered Bravais lattices, do not have the point symmetry of the lattice. However, each Bravais lattice has a primitive unit cell that has the point symmetry of the lattice. This cell is called the Wigner–Seitz cell.

2.14.2 *Construction of the Wigner–Seitz Cell*

The Wigner–Seitz cell, like every primitive unit cell, contains only one lattice point, but this point has a very particular location in the cell. It is placed in the geometric center of the cell and the region of space that is closer to that point than to any other lattice point defines this cell. In order to obtain the Wigner–Seitz cell, we have to identify first the NNs of a lattice point. The NNNs may also be involved in the construction of that cell and even the TNNs. This cell can be obtained in the following manner:

(a) First, any point of the lattice is chosen (the one that is going to be in the middle of the Wigner–Seitz cell).
(b) Second, we connect this lattice point with all the NNs by means of segments and draw median planes of the segments. In this manner a three-dimensional body, limited by these planes, is obtained.
(c) Last, we repeat the same work as in point (b), but with the NNNs. If the new planes reduce the volume of the region defined by the first planes, this new volume will be the Wigner–Seitz cell, if, of course, more distant neighbors (TNNs, fourth nearest neighbors, and so on) do not manage to limit this volume even more.

2.14.3 *The Wigner–Seitz Cell of the bcc Lattice*

Figure 2.42 shows the Wigner–Seitz cell of the *bcc* lattice. This cell has the shape of a tetradecahedron (a polyhedron with 14 faces). Eight of its faces are defined by 8 NNs and the rest of them by 6 NNNs. This tetradecahedron may be seen as a truncated regular octahedron. That is, the faces of the octahedron, which are defined

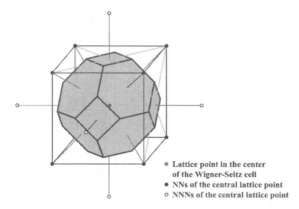

• Lattice point in the center
  of the Wigner–Seitz cell
• NNs of the central lattice point
○ NNNs of the central lattice point

Figure 2.42 The Wigner–Seitz cell of the *bcc* lattice.

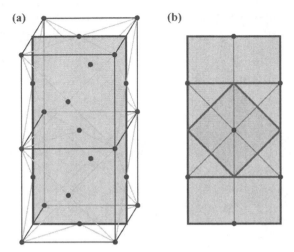

Figure 2.43   In (a) is outlined a cross section of two cubic F cells of the *fcc* lattice. In (b) the cross section is used to demonstration that in the construction of the Wigner–Seitz cell participate only the NNs of the lattice point belonging to this cell.

by 8 NNs, are truncated by the 6 faces defined by the NNNs. This truncated octahedron has 6 square faces at a distance of $a/2$ (where a is the edge length of the cubic I cell) from the middle of the Wigner–Seitz cell and 8 hexagonal faces at a distance of $(\sqrt{3}/4)a$ from the center. It is easy to identify, by looking at the number, shape, and orientation of the faces of the truncated octahedron, the 4 threefold axes and 3 fourfold axes that the cubic I unit cell of the *bcc* lattice has (see Fig. 2.42).

2.14.4 *The Wigner–Seitz Cell of the fcc Lattice*

Figure 2.43a shows a cross section of two cubic F unit cells of the *fcc* lattice, in which there are 7 lattice points: the central one, 4 of its NNs, and 2 NNNs. The smaller square in Fig. 2.43b represents a cross section of the Wigner–Seitz cell. We demonstrate in this figure that in the construction of the Wigner–Seitz cell participate only the nearest neighboring lattice points. Since the number of the NNs in the *fcc* lattice is 12, its Wigner–Seitz cell has the shape of a dodecahedron. This is a rhombic dodecahedron, and it is shown in

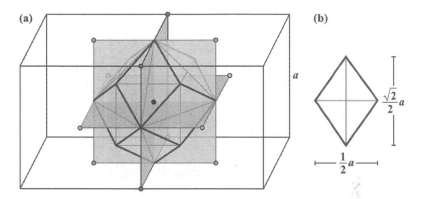

Figure 2.44 (a) The Wigner–Seitz cell of the *fcc* lattice. (b) A face of the dodecahedron shown in (a).

Fig. 2.44a. Figure 2.44b displays one of the 12 identical faces of this dodecahedron.

Problems

Exercise 2.1 In Fig. 2.2b are shown two twofold and one fourfold rotation axes of a square prism. Are there any other twofold rotation axes? Draw them, if any. Justify your answer.

Exercise 2.2 Figures 2.45a and 2.45b depict a regular tetrahedron and octahedron, respectively.

(a) Draw all the rotation axes of the regular tetrahedron.
(b) Do the same as in (a) for the regular octahedron.

Exercise 2.3 We can appreciate in Fig. 2.46 a truncated regular octahedron inscribed in a cube of edge length a, while the regular octahedron by itself is inscribed in a cube of edge length $(3/2)a$. The faces of the two cubes are parallel to each other and their geometric centers coincide. It is shown in this figure how the octahedron is truncated by the faces of the smaller cube with a tetradecahedron as a result. This tetradecahedron has 8 faces in shape of a regular hexagon (can you explain why?) and 6 faces in shape of a square.

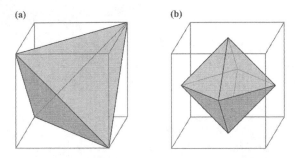

Figure 2.45   (a) A regular tetrahedron and (b) a regular octahedron.

(a) Draw all the rotation axes of the tetradecahedron (or truncated octahedron) from Fig. 2.46.

(b) Explain why this tetradecahedron does not have sixfold rotation axes.

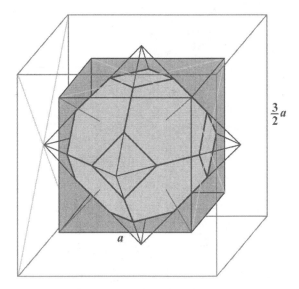

Figure 2.46   A tetradecahedron inscribed in a cube of edge length a. This tetradecahedron may be seen as a regular octahedron truncated by 6 faces of the smaller cube. The regular octahedron is inscribed in a cube of edge length $(3/2)a$ and has the same geometric center as the cube of edge length a.

Exercise 2.4 Figure 2.47 shows the Wigner–Seitz cell of the *fcc* lattice, which has the shape of a rhombic dodecahedron.

(a) Draw the fourfold rotation axes of the rhombic dodecahedron. How many such axes does it have?
(b) Repeat (a) for threefold rotation axes.

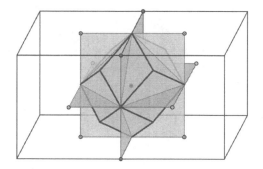

Figure 2.47 The Wigner–Seitz cell of the *fcc* lattice.

Exercise 2.5 In Fig. 2.48, it is defined a set of 27 points. Show that this set of points has threefold and fourfold rotation axes. How many axes of each type does it have?

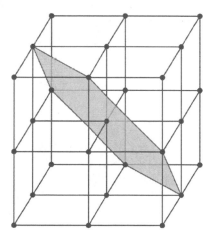

Figure 2.48 A set of 27 points located at the vertices of the 8 small cubes.

Exercise 2.6 How many twofold rotation axes does the set of 27 points defined in Fig. 2.48 possess? Draw them.

Exercise 2.7 Show that the monoclinic lattices from Figs. 2.49a and 2.49b are equivalent.

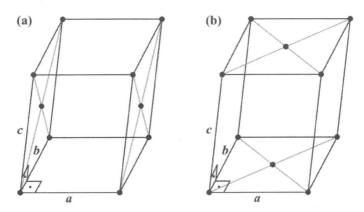

Figure 2.49 Monoclinic cells centered in two different ways: (a) *A*-face centered and (b) *C*-face centered. The *b*-axis setting is assumed.

Exercise 2.8 Find graphically the NNs, NNNs, TNNs, and also fourth nearest neighbors of a given lattice point in a two-dimensional hexagonal lattice of lattice constant a. Calculate their distances to the lattice point in consideration. Why the number of NNs, NNNs, and so on is a multiple of 6?

Exercise 2.9 Let us consider a two-dimensional centered rectangular lattice.

(a) Verify that, when the length of edges a, b of the conventional rectangular unit cell fulfill the relation $b/a = 1/2$, the NNs, NNNs, TNNs, and fourth nearest neighbors of a given lattice point are at distances

$$\frac{\sqrt{4}}{4}a, \ \frac{\sqrt{5}}{4}a, \ \frac{\sqrt{13}}{4}a, \ \text{and} \ \frac{\sqrt{16}}{4}a$$

from this point, respectively. Show these neighbors graphically.

(b) Repeat (a) for the case of $b/a = 2/3$. Then, the NNs, NNNs, TNNs, and fourth nearest neighbors of a lattice point are at distances

$$\frac{\sqrt{13}}{6}a, \quad \frac{\sqrt{16}}{6}a, \quad \frac{\sqrt{36}}{6}a, \quad \text{and} \quad \frac{\sqrt{45}}{6}a$$

from this point, respectively.

Exercise 2.10 In Fig. 2.50, we can find all the NNs, some of the NNNs, and also some of the TNNs of a lattice point located at the center of the displayed *fcc* lattice with a lattice constant a.

(a) Show that the NNs, NNNs, TNNs, and also fourth and fifth nearest neighbors of a lattice point in the *fcc* lattice are at distances

$$\frac{\sqrt{2}}{2}a, \quad \frac{\sqrt{4}}{2}a, \quad \frac{\sqrt{6}}{2}a, \quad \frac{\sqrt{8}}{2}a, \quad \text{and} \quad \frac{\sqrt{10}}{2}a$$

from this point, respectively.

(b) Place all the NNNs and the TNNs of the lattice point in consideration that fit in the empty cubes shown in Fig. 2.50.

(c) Is it possible to estimate the number of NNNs and TNNs of a lattice point in the *fcc* lattice using the information obtained in point (b)?

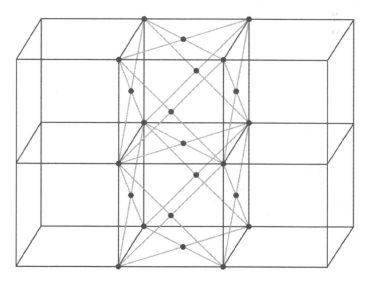

Figure 2.50 All the NNs, some of the NNNs, and also some of the TNNs of a lattice point located at the center of the displayed *fcc* lattice.

Exercise 2.11 Figure 2.51 shows a two-dimensional square lattice of lattice constant a. The coordinate doublet of the lattice point (in units of a) is given by a set of two integer numbers n_1, n_2. Let us consider all the neighbors of the lattice point placed at the origin **O**, whose distances d to it fulfill the relation $d < 6a$. Notice that the origin represents a fourfold rotation point of the lattice and only $1/4$ of the region to which belong such points is shown in Fig. 2.51 (remember, that the points from the limit of the region belong in $1/2$ to it).

(a) Nearby to each lattice point from the highlighted region, write its distance to the origin **O** in units of a. For example, in the case of the lattice point specified in Fig. 2.51 ($n_1 = 3$, $n_2 = 2$) this distance is $\sqrt{13}$.

(b) Calculate the numbers of lattice points that are at the same distance from the point placed at the origin.

(c) Can you explain why are there 12 thirteenth nearest neighbors of a given lattice point and only 4 or 8 in the other cases considered here?

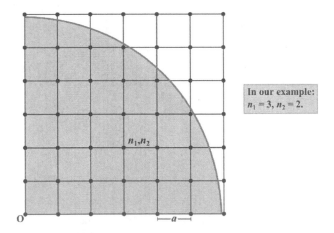

Figure 2.51 A two-dimensional square lattice of lattice constant a.

Exercise 2.12 The *sc* lattice of lattice constant a may be considered as built up of two-dimensional square lattices of the same lattice constant, stacked one on top of the other. The set of three numbers, n_1, n_2, n_3, where $n_1, n_2, n_3 \in \mathbb{Z}$, represents the coordinate triplets of a lattice point in units of a. The points of coordinate triplets $n_1, n_2, 0$ belong to the square lattice shown in Fig. 2.51, while the points identified by $n_1, n_2, \pm1; n_1, n_2, \pm2; n_1, n_2, \pm3$ belong to the square lattices that are at distances a, $2a$, and $3a$ from the first one, respectively. Let us now consider the lattice points whose distances, d, to the origin $\mathbf{O}$, fulfill the relation

$$d = a\sqrt{n_1^2 + n_2^2 + n_3^2} \leqslant a\sqrt{10}. \tag{2.10}$$

In the square lattice shown in Fig. 2.51, we have 36 such lattice points at the following distances from the origin

$$a\sqrt{1},\ a\sqrt{2},\ a\sqrt{4},\ a\sqrt{5},\ a\sqrt{8},\ a\sqrt{9},\ a\sqrt{10}$$

and the number of such points is 4, 4, 4, 8, 4, 4, and 8, respectively.

(a) Find all the distances of the points indexed by $n_1, n_2, \pm1$ that fulfill Eq. 2.10 and the number of neighbors at each distance from the origin. Do the same for the lattice points indexed by $n_1, n_2, \pm2$ and $n_1, n_2, \pm3$.
(b) Are there any lattice points at a distance of $a\sqrt{7}$ from the origin? If not, why not?
(c) Calculate the number of NNs, NNNs, TNNs, and so on up to ninth nearest neighbors of a given lattice point of the sc lattice.

Exercise 2.13 Choose one of the 12 NNNs of the central lattice point of the sc lattice shown in Fig. 2.24. Find the rest of the NNNs by doing rotations. Describe your choice of rotations that you have done in order to obtain the remaining 11 NNNs. Also, draw the axes that you have used, and explain in what order the rotations have to be done.

Exercise 2.14 Figure 2.52 shows a rhombohedron constructed inside a hexagonal prism of side a and height c. Two vertices of the rhombohedron are located at the geometric centers of the hexagonal bases and the other 6 form two groups of 3 vertices each. The location of the vertices belonging to each group is described in Section 2.6. Show that for $c/a = \sqrt{6}/2$ the rhombohedron takes on the shape of a cube.

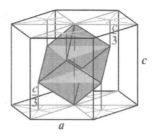

Figure 2.52   A rhombohedron constructed inside a hexagonal prism.

Exercise 2.15   The set of points from Fig. 2.48 defines a *sc* lattice.

(a) Draw two replicas of the regular hexagon depicted there in such a way that the geometric centers of the new hexagons coincide with the end points of the large cube body diagonal (orthogonal to the planes of the hexagons). Next, trace the edges orthogonal to the bases in order to obtain two identical hexagonal prisms that share a common base. Each hexagonal prism contains the primitive cubic unit cell of the *sc* lattice in consideration.

(b) Find the NNs of the lattice point located at the geometric center of the large cube and draw plane figures defined by them in planes parallel to the bases of the hexagonal prisms.

Exercise 2.16   Let us consider a *fcc* lattice of a lattice constant a_c.

(a) Draw a hexagonal prism and a rhombohedral P unit cell inscribed in it for the *fcc* lattice (see Fig. 2.53a). Have in mind that the side a_h and height c_h of this hexagonal prism fulfill the relation $c_h/a_h = \sqrt{6}$ (see Table 2.7).

(b) Find the NNs of the lattice point located at the center of the bottom base of the hexagonal prism (those that are included in your draw).

(c) Where are the rest of the NNs of the lattice point in consideration?

Exercise 2.17   Give the position vectors of all the NNs of the lattice point placed at the origin **O** of the three *bcc* unit cells (cubic I, rhombohedral P, and triple hexagonal R) drawn in Fig. 2.41. These

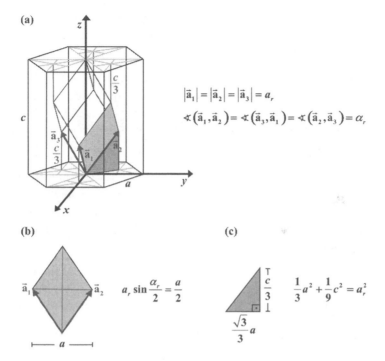

(a)

$$|\vec{a}_1| = |\vec{a}_2| = |\vec{a}_3| = a_r,$$

$$\sphericalangle(\vec{a}_1,\vec{a}_2) = \sphericalangle(\vec{a}_3,\vec{a}_1) = \sphericalangle(\vec{a}_2,\vec{a}_3) = \alpha_r$$

(b)

$$a_r \sin\frac{\alpha_r}{2} = \frac{a}{2}$$

(c)

$$\frac{1}{3}a^2 + \frac{1}{9}c^2 = a_r^2$$

Figure 2.53 (a) A rhombohedron constructed inside a hexagonal prism of side a and height c. (b) The highlighted face of the rhombohedron from (a). (c) The right triangle highlighted in (a).

vectors should be expressed in terms of the basis vectors $\vec{a}_r$, $\vec{b}_r$, and $\vec{c}_r$ defined in the figure.

Exercise 2.18 A lattice point of the *bcc* lattice has 6 NNNs, 12 TNNs, and 24 fourth nearest neighbors. In Fig. 2.41, you can find some of these neighbors of the lattice point placed at the common origin **O** of the three unit cells (cubic I, rhombohedral P, and triple hexagonal R) shown in the figure.

(a) How many NNNs of the lattice point placed at the origin **O** is included in the draw? Where they are located? Give their position vectors expressed in terms of the convenient set of basis vectors that are defined in the figure.
(b) Repeat (a) for the TNNs and fourth nearest neighbors.

Exercise 2.19 In Fig. 2.53a it is shown a rhombohedron constructed inside a hexagonal prism of side a and height c. The rhombohedron is defined by the vectors $\vec{a}_1$, $\vec{a}_2$, and $\vec{a}_3$.

(a) Show that the vector given by the sum $(\vec{a}_1 + \vec{a}_2 + \vec{a}_3)$ is lying along the longest diagonal of the rhombohedron and its longitude is c. Find the sum of the vectors graphically.

(b) Calculate the volume of the rhombohedron and compare it with the volume of the hexagonal prism.

Exercise 2.20 Figure 2.53b shows the face of the rhombohedron that is highlighted in Fig. 2.53a. This face is defined by vectors $\vec{a}_1$ and $\vec{a}_2$. Figure 2.53c, in turn, displays the right triangle highlighted in Fig. 2.53a. This triangle is defined by vector $\vec{a}_2$ and its projection onto the bottom base of the hexagonal prism.

(a) Using the plane figures from Figs. 2.53b and 2.53c, show that the relation between the parameters a_r and α_r that describe the rhombohedron and the parameters a and c that describe the hexagonal prism (in which this rhombohedron is inscribed) is the following

$$\begin{cases} a_r = \dfrac{1}{3}\sqrt{3a^2 + c^2} = \dfrac{a}{3}\sqrt{3 + (c/a)^2} \\[2mm] \sin\dfrac{\alpha_r}{2} = \dfrac{3}{2\sqrt{3 + (c/a)^2}} \end{cases} \qquad (2.11)$$

(b) Show that the c/a ratio is expressed only by the parameter α_r of the rhombohedron

$$\frac{c}{a} = \sqrt{\frac{9}{4\sin^2(\alpha_r/2)} - 3}. \qquad (2.12)$$

Exercise 2.21 Using Eq. 2.12 verify the following relations

$$\begin{cases} (c/a)_{fcc} = \sqrt{6} \\[2mm] (c/a)_{sc} = \dfrac{1}{2}(c/a)_{fcc} \\[2mm] (c/a)_{bcc} = \dfrac{1}{4}(c/a)_{fcc} \end{cases} \qquad (2.13)$$

where a and c correspond to the side and height of the hexagonal prism, respectively, in which is inscribed (see Fig. 2.53) the primitive unit cell of a cubic lattice (sc, bcc, or fcc).

Exercise 2.22 We know from Table 2.5 that mercury in the α phase, α-Hg, crystallizes in the crystal structure with rhombohedral space-group symmetry. Its rhombohedral P unit cell contains one atom.

(a) Draw in real proportions the hexagonal prism composed of three triple hexagonal R unit cells for the rhombohedral lattice of the α-Hg structure at 227 K and normal pressure, and attach the atomic basis to each lattice point. For that purpose, calculate first the c/a ratio using Eq. 2.12 and the data provided in Table 2.5.
(b) Find the NNs of an atom placed in the center of the bottom base of the prism. What percentage of NNs is included in your draw? What is the distance between an atom to its NNs?

Exercise 2.23 Calculate the angles between the axes that define the primitive rhombohedral unit cell of the *bcc* lattice.

Exercise 2.24 Figure 2.54 shows a body centered rhombohedral unit cell. To which Bravais lattice does belong this cell?

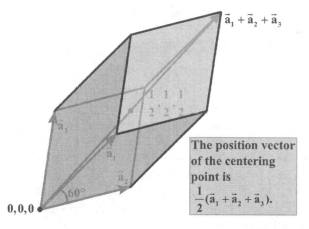

Figure 2.54 Body centered rhombohedral unit cell. The coordinates of the two lattice points are expressed in units of a_r ($a_r = |\vec{a}_1| = |\vec{a}_2| = |\vec{a}_3|$).

Exercise 2.25 The *bcc* lattice can be built by placing an additional point in the geometric center of each small cube in Fig. 2.34. The unit cell, defined by basis vectors $\vec{a}_1, \vec{a}_2, \vec{a}_3$ in this figure (with edges at an angle of 60° to each other), represents a centered rhombohedral unit cell of the obtained *bcc* lattice.

(a) How many *bcc* lattice points do belong to such rhombohedral unit cell?
(b) Give the coordinate triplets (with respect to basis vectors $\vec{a}_1, \vec{a}_2, \vec{a}_3$) of the lattice points belonging to the rhombohedral unit cell in consideration.

Exercise 2.26 For the *sc* lattice of lattice constant *a*, draw the smallest rhombohedral unit cell defined by basis vectors with angles of 109°28′ between them.

(a) What type of rhombohedral unit cell did you obtain?
(b) How many lattice points do belong to this cell?
(c) Calculate the volume of the rhombohedral unit cell and compare it with the volume of the primitive cubic cell.

Exercise 2.27 A rhombohedron with edges at an angle of 109°28′ to each other represents a centered unit cell not only for the *sc* lattice but also for the *fcc* lattice. Figure 2.55 shows such rhombohedron. Its shortest diagonal coincides with a body diagonal of the cube that represents the cubic *F* unit cell of the *fcc* lattice.

(a) How many *fcc* lattice points do belong to the centered rhombohedral unit cell in consideration?
(b) Describe, how the cubic *F* unit cell of the *fcc* lattice is inscribed in the centered rhombohedral unit cell in consideration.

Exercise 2.28 Similarly as it was done in Fig. 2.41 for the *bcc* lattice, draw the hexagonal prism composed of the three triple hexagonal *R* unit cells for the *fcc* lattice. The resulting picture should contain the cubic *F*, rhombohedral *P*, and triple hexagonal *R* unit cells for the *fcc* lattice, all of them with a common origin.

Exercise 2.29 In a hexagonal lattice, draw a *C*-centered orthorhombic unit cell.

Exercise 2.30 Let us consider a centered orthorhombic unit cell in a rhombohedral lattice.

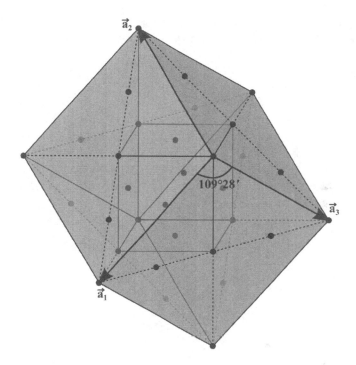

Figure 2.55 A centered rhombohedral unit cell for the *fcc* lattice.

(a) In Fig. 2.38b, draw the projection of a centered orthorhombic unit cell present in a rhombohedral lattice. Locate in your draw the basis vectors $\vec{a}_{ortho}$ and $\vec{b}_{ortho}$ that define the basis of this cell.

(b) How many centering points does this cell have?

(c) Give the coordinate triplets of the centering points in terms of the basis vectors $\vec{a}_{ortho}$, $\vec{b}_{ortho}$, $\vec{c}_{ortho}$ of the centered orthorhombic unit cell.

Exercise 2.31 Let us consider a Wigner–Seitz cell of a two-dimensional lattice. Then a median plane of the segment which connects a lattice point with its NNs becomes a straight line.

(a) Construct the Wigner–Seitz cell of a two-dimensional hexagonal lattice. Which neighbors do participate in this construction?

(b) Repeat (a) for the two-dimensional centered rectangular lattices defined in Exercise 2.9.

Exercise 2.32 Construct the Wigner–Seitz cell of a three-dimensional hexagonal lattice. Which neighbors do participate in this construction?

Exercise 2.33 When we draw median planes of the segments which connect a lattice point of the *bcc* lattice with its NNs, a regular octahedron is obtained. In order to obtain the Wigner–Seitz cell, this work has to be repeated with the NNNs and the truncated regular octahedron depicted in Fig. 2.42 is the result. Figure 2.46 shows both the octahedron and the resulting truncated octahedron. The latter is inscribed in the cubic unit cell of the *bcc* lattice, and the former is inscribed in a larger cube. The centers of the two cubes coincide and their faces are parallel to each other. Show that the larger cube is of edge length $(3/2)a$, where a is the edge length of the cubic unit cell of the *bcc* lattice.

Chapter 3

Crystal Structures of Elements

3.1 Introduction

In this chapter, we will consider the crystal structure of most metallic elements, nonmetals from column IV of the periodic table, and noble gases. Arsenic and selenium from columns V and VI, respectively, are also included in our analysis. More than 30 elements crystallize in two monoatomic crystal structures, *fcc* and *bcc*, at room temperature and normal pressure. There are also a large number of elements that crystallize in a structure that can be described by a hexagonal Bravais lattice but with two-atom basis. This is the so-called *hexagonal close-packed* (*hcp*) crystal structure. To describe this structure, identical spheres are arranged in a regular array to minimize the interstitial volume. This close-packing of spheres may lead, however, to many different arrangements. One of them turns out to be nothing more than the *fcc* structure. Under normal conditions, more than 40% of the elements crystallize in the *hcp*, *fcc*, and other close-packed crystal structures. All of them will be considered in this chapter.

In some sense, the idea of close-packing of spheres, to obtain crystal structures, coincides with the idea to consider atoms (or ions) as impenetrable hard spheres of a certain radius r. This model,

Basic Elements of Crystallography (2nd Edition)
Nevill Gonzalez Szwacki and Teresa Szwacka
Copyright © 2016 Pan Stanford Publishing Pte. Ltd.
ISBN 978-981-4613-57-6 (Hardcover), 978-981-4613-58-3 (eBook)
www.panstanford.com

even being so simple, is quite useful in the description of crystal structures. For example, it allows for the prediction of interatomic distances of new structures to a first order approximation. The atomic radius is deduced from observed atomic separations in a set of crystals. However, the results may vary from set to set since the atomic separation depends on the type of chemical bonding. (The three principal types of chemical bonds in crystals are ionic, covalent, and metallic.) The radius of an atom in the crystal of an element is given by half of the observed minimal atomic separation.

The atomic radius depends on the type of bond in the crystal because the nature of bonding is strongly connected to the spatial distribution of electrons. The degree of impenetrability of atoms (or ions) depends on their electronic configuration. The highest impenetrability is achieved in the case of atoms (or ions) with closed electron shells. This is, e.g., the case of noble gases, positive ions of alkali metals (Li^+, Na^+, K^+, Rb^+, or Cs^+) or negative ions of the halogens (F^-, Cl^-, Br^-, or I^-). The high degree of impenetrability of such an atom (or ion) is a consequence of the Pauli exclusion principle and a large energy gap existing between the lowest unoccupied atomic orbital and the highest occupied one.

The idea to consider an atom (or ion) as a hard sphere will be used frequently in this chapter.

3.2 Pearson Notation and Prototype Structure

The *Pearson notation*, together with the prototype structure, allows shorthand characterization of crystal structures. It consists of the symbol of the Bravais lattice corresponding to the structure in consideration followed by the number of atoms per conventional unit cell. Table 3.1 lists *Pearson symbols* for the 14 Bravais lattices. The assignation of the Pearson symbol to a crystal structure is not unique; it means that in general one Pearson symbol corresponds to more than one crystal structure. To achieve a unique identification of a crystal structure, a representative (prototype) element or compound (in a proper phase) having that structure is assigned to each structure type. The Pearson symbol together with the prototype structure identifies the crystal structure of a given

Table 3.1 Pearson symbols corresponding to 14 Bravais lattices. In the symbols, *n* expresses the number of atoms per conventional unit cell. The last column gives examples of Pearson symbols, which together with the prototype structures correspond to crystal structures of elements

Crystal system	Bravais lattice symbol	Pearson symbol	Examples of crystal structures
Triclinic (anorthic)	*aP*	*aPn*	
Monoclinic	*mP*	*mPn*	*mP*4-Bi
	mS (*mA, mB, mC*)	*mSn*	*mS*4-Bi
Orthorombic	*oP*	*oPn*	*oP*8-Np
	oS (*oA, oB, oC*)	*oSn*	*oS*4-U
	oI	*oIn*	
	oF	*oFn*	*oF*8-Pu
Tetragonal	*tP*	*tPn*	*tP*4-Np
	tI	*tIn*	*tI*2-In
Trigonal	*hP*	*hPn*	*hP*3-Se
	hR	*hRn**	*hR*1-Hg
			*hR*1-Po
			*hR*3-Sm
Hexagonal	*hP*	*hPn*	*hP*2-Mg
			*hP*4-La
			*hP*4-C
Cubic	*cP*	*cPn*	*cP*1-Po
	cI	*cIn*	*cI*2-W
	cF	*cFn*	*cF*4-Cu

*In this symbol, *n* refers to the primitive rhombohedral unit cell.

element or compound. The examples listed in the last column of Table 3.1 correspond to crystal structures of elements.

3.3 Filling Factor

The *filling factor* of a crystal structure is defined as the fraction of the total crystal volume filled with atoms considered as hard spheres. Sometimes instead of filling factor the expressions *atomic packing factor* or *packing fraction* are used. The filling factor gives us an idea how close the atoms are "packed" in the crystal structure. The closest packing of atoms is achieved when the number of NNs is the largest possible. Therefore, the filling factor combined with the coordination number gives us an idea about the degree of filling of the crystal volume with atoms and, at the same time, tells us how close the atoms in a crystal are packed.

In order to calculate the filling factor, we have to know first the radii for atoms considered as hard spheres. In the case of elements, the radius is half the distance between NNs. To do the calculations it is enough to consider the volume of a unit cell of the crystal structure. The filling factor is defined as

$$filling\ factor = \frac{volume\ occupied\ by\ atoms\ (hard\ spheres)\ within\ the\ unit\ cell}{cell\ volume}. \tag{3.1}$$

In the following, we will learn about the different crystal structures in which crystallize the elements starting with the *sc* structure.

3.4 Simple Cubic Structure

Pearson symbol: *cP1*, prototype: Po. Under normal conditions, only one element, polonium in the α phase, crystallizes in the *sc* structure. However, there are three elements, As, Sb, and Bi (discussed already in the previous chapter), that crystallize in the *hR2*-As structure, which is a slightly distorted *sc* structure.

Let us assume that the vertices of the cubic P unit cell of the *sc* structure coincide with the centers of atoms. Figure 3.1 shows the plane of one of the faces of the cube with the cross sections of the atoms that, being considered as hard spheres, are represented by circles on this plane. Each atom from a cube face has two of its NNs on this face. The NNs are at a distance equal to the lattice constant a. The atomic radius is equal to half of a, so the volume of the only one atom belonging to the cube is given by

$$V_{atom} = \frac{4}{3}\pi \left(\frac{a}{2}\right)^3 = \frac{\pi}{6}a^3 \tag{3.2}$$

and the filling factor for the *sc* structure is

$$(filling\ factor)_{sc} = \frac{\frac{\pi}{6}a^3}{a^3} = \frac{\pi}{6} \cong 0.52. \tag{3.3}$$

The result for the filling factor shows that in the case of the *sc* structure about half of the crystal volume is filled with atoms and

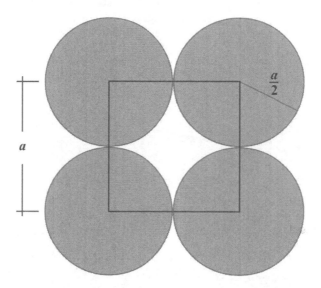

Figure 3.1 The plane that contains a face of the cubic P unit cell of the sc structure with the cross sections of atoms considered as hard spheres.

the other half corresponds to the interstices. Thus, the interstitial volume in the sc structure is quite large. Of course, this is reflected also by the coordination number; the number of the NNs of an atom in the case of the sc structure is only 6. The NNNs are already 12, but at a distance about 40% larger than the distance to the NNs.

3.5 Body Centered Cubic Structure

Pearson symbol: $cI2$, prototype: W. Now, we will see the case of the bcc structure. In this case, the atoms of the vertices of the cube are NNs of the atom that is in the center of the cube (see Fig. 3.2a), and those atoms are in contact with it. The point of contact between two atoms is found in a plane defined by two body diagonals of the cube, as shown in Fig. 3.2. We can see in Fig. 3.2b that the atoms are in contact with each other only along the body diagonals of the cube, while the atoms that are at the vertices are at a distance greater than $2r$ (r – radius of the atom). There are two atoms in the cubic I cell of

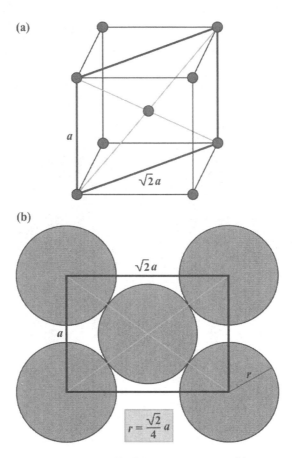

Figure 3.2 (a) Cubic *I* unit cell of the *bcc* structure. (b) Cross sections of atoms considered as hard spheres in a plane defined by two body diagonals of the cube shown in (a). In this plane, there are the points of contact between the central atom and its NNs.

the *bcc* structure, so the filling factor is

$$(\text{filling factor})_{bcc} = \frac{2\frac{4}{3}\pi \left(\frac{\sqrt{3}}{4}a\right)^3}{a^3} = \frac{\sqrt{3}}{8}\pi \cong 0.68. \qquad (3.4)$$

Note that the filling factor for the *bcc* structure is higher than that one for the *sc* structure. This is consistent with the fact that the number of NNs in *bcc* is also larger than in *sc* (8 and 6, respectively). Moreover, the distance of the 6 NNNs of an atom in the *bcc* structure

Table 3.2 Lattice constants of elements that crystallize in the *bcc* structure at normal pressure. The data is provided at room temperature, unless otherwise specified

Metal	$a(\text{Å})$	Metal	$a(\text{Å})$
α-Ba	5.023	β-Pm	4.100 (1163 K)
β-Ca	4.380 (773 K)	β-Pr	4.130 (1094 K)
δ-Ce	4.120 (1030 K)	ε-Pu	3.638 (773 K)
α-Cr	2.8847	Ra	5.148
α-Cs	6.141	α-Rb	5.705
β-Dy	4.030 (1654 K)	β-Sc	3.752 (1623 K)
α-Eu	4.5827	β-Sr	4.850 (887 K)
α-Fe	2.8665	Ta	3.3031
δ-Fe	2.9346 (1712 K)	β-Tb	4.070 (1562 K)
β-Gd	4.060 (1538 K)	β-Th	4.110 (1723 K)
K	5.321	β-Ti	3.3065 (1173 K)
γ-La	4.260 (1160 K)	β-Tl	3.882 (506 K)
β-Li	3.5093	γ-U	3.524 (1078 K)
δ-Mn	3.081 (1413 K)	V	3.024
Mo	3.147	W	3.1651
β-Na	4.291	β-Y	4.100 (1751 K)
Nb	3.3007	γ-Yb	4.440 (1036 K)
β-Nd	4.130 (1156 K)	β-Zr	3.609 (1135 K)
γ-Np	3.520 (873 K)		

differs from the distance of its NNs by less than 15%. Therefore, an atom in this structure has effectively 14 atoms close to it.

Table 3.2 reports lattice constants a for all elements that crystallize in the *bcc* structure at room temperature and normal pressure. All of them are metals. In this table, we can also find lattice constants for a number of metals that crystallize in the *bcc* structure at high temperatures and normal pressure. Under normal conditions, these metals (with the exception of Fe) crystallize in structures different from *bcc* and they will be considered later. One of them, manganese (α-Mn), crystallizes in a very complex structure due to its magnetic (antiferromagnetic) properties. This structure may be considered as *bcc* with 56 additional atoms, which means in total 58 atoms per unit cell (Pearson symbol *cI*58). On the other hand, this supercell may be viewed as built of $3 \times 3 \times 3 = 27$ cubic *I* unit cells containing $2 \times 3 \times 3 \times 3 = 54$ atoms with still 4 additional

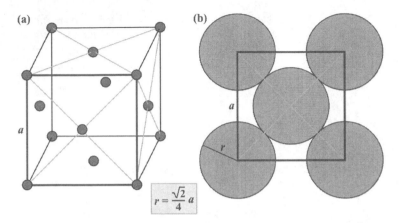

Figure 3.3 (a) Cubic F unit cell of the *fcc* structure. (b) Cross sections of 5 atoms considered as hard spheres from the front face of the cube from (a). The points of contact between the NNs are found in this plane.

atoms added. A number of these atoms are slightly shifted from the ideal positions in the small cubic I unit cells. The NN interatomic distances in α-Mn, with lattice constant $a = 8.9125$ Å (at 298 K), are in the range of $2.244 - 2.911$ Å.

3.6 Face Centered Cubic Structure

Pearson symbol: $cF4$, prototype: Cu. We will now turn to the case of the *fcc* structure. The cubic F unit cell of this structure is shown in Fig. 3.3a. The atom placed in the center of a face of the cube has 4 of its NNs at the vertices of this face. Figure 3.3b shows the plane of the front face of the cube with the cross sections of 5 atoms considered as hard spheres. The points of contact between the atoms are found on the face diagonals. In the *fcc* structure, there are four atoms in the cubic unit cell; therefore the filling factor is

$$(\textit{filling factor})_{fcc} = \frac{4\frac{4}{3}\pi \left(\frac{\sqrt{2}}{4}a\right)^3}{a^3} = \frac{\sqrt{2}}{6}\pi \cong 0.74. \quad (3.5)$$

In this case $3/4$ of the crystal volume is filled with atoms considered as hard spheres and only $1/4$ is empty. The number of the NNs, equal to 12, is also the largest possible. This filling factor is the largest one

Table 3.3 Lattice constants of elements that crystallize in the *fcc* structure at normal pressure. The data is given at room temperature, unless otherwise specified

Element	$a(\text{Å})$	Element	$a(\text{Å})$
Ac	5.311	β-La	5.303 (598 K)
Ag	4.0861	γ-Mn	3.863 (1373 K)
α-Al	4.0496	Ne	4.462 (4.2 K)
Ar	5.311 (4.2 K)	Ni	3.5241
Au	4.0784	α-Pb	4.9502
α-Ca	5.5884	Pd	3.8901
α-Ce	4.850 (77 K)	Pt	3.924
γ-Ce	5.1610	δ-Pu	4.637 (592 K)
α-Co	3.569 (793 K)	Rh	3.803
Cu	3.6149	α-Sr	6.084
Es	5.750	α-Th	5.084
γ-Fe	3.630 (1373 K)	Xe	6.309 (145 K)
Ir	3.8391	β-Yb	5.4848
Kr	5.796 (96 K)		

among the filling factors for the cubic structures and at the same time the larges one among the filling factors for all the structures of elements.

Table 3.3 lists lattice constants a of all metals that crystallize at room temperature and normal pressure in the *fcc* structure. Besides that, in this table are also given the lattice constants of four noble gases (argon, krypton, neon, and xenon) and a number of metals that crystallize in the *fcc* structure at temperatures different from room temperature. A similar number of metallic elements crystallize in the *fcc* and *bcc* structures under normal conditions, as can be seen by comparing Tables 3.2 and 3.3.

The *fcc* structure represents one of the close-packed structures. We will discuss them below.

3.7 Close-Packed Structures

The name "close-packed" refers to the way of packing the atoms in order to obtain the highest possible filling factor. To consider close-packed structures it is worthwhile to analyze first the manner

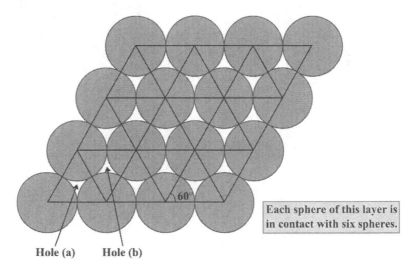

Each sphere of this layer is in contact with six spheres.

Hole (a)   Hole (b)

Figure 3.4 A close-packed layer of equal spheres that is a two-dimensional close-packed hexagonal structure.

of placing spheres of the same radius, in order for the interstitial volume to be as small as possible. For this purpose, the spheres are arranged in layers that are placed one on top of the other in the way we will explain later here.

Each sphere within a layer is in contact with six others, and the layer itself represents a two-dimensional close-packed hexagonal structure. The cross section of a layer is shown in Fig. 3.4. We will differentiate the holes existing between spheres of a layer as of type (a) or type (b) (see Fig. 3.4). Figure 3.5 shows the plane defined by the centers of spheres of the first layer and the projection of the centers of spheres of the second layer. The centers of spheres of the second layer are above the centers of the holes of type (a) specified in Fig. 3.4. The spheres of the second layer just rest in the holes of type (a). The centers of the holes of type (a) coincide with the geometric centers of the equilateral triangles shown in Fig. 3.5 (of course the same occurs in the case of holes of type (b), see Fig. 3.6a). Therefore, each sphere of the second layer is in contact with three spheres of the layer below it. A third layer can be placed in two ways as depicted in Fig. 3.6. In the case shown in Fig. 3.6a, the centers of spheres of the third layer are above the centers of the holes of

Center of the first layer sphere

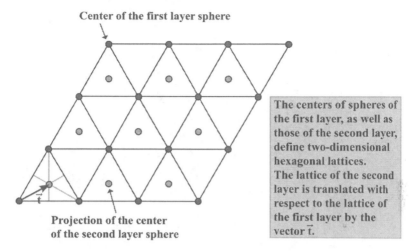

The centers of spheres of
the first layer, as well as
those of the second layer,
define two-dimensional
hexagonal lattices.
The lattice of the second
layer is translated with
respect to the lattice of
the first layer by the
vector $\vec{t}$.

**Projection of the center
of the second layer sphere**

Figure 3.5 The centers of spheres of the first layer and the projection of
the centers of spheres of the second layer in a close-packed arrangement of
equal spheres.

type (b) of the first layer, specified in Fig. 3.4, whereas in the case
shown in Fig. 3.6b, the spheres of the third layer lie directly above
the spheres of the first layer.

We will show now that the close-packed arrangement displayed
in Fig. 3.6a corresponds to the *fcc* structure. A part of Fig. 3.6a, with
the cubic cell of the *fcc* structure, is drawn in Fig. 3.7. We can see
in this figure that the *fcc* structure is of **ABCABC...** type, where **A**,
B, and **C** denote three two-dimensional close-packed layers shifted
horizontally one with respect to the other. The layer planes are
orthogonal to a body diagonal of the cubic unit cell of this structure.
The second layer, **B**, is shifted with respect to the first one, **A**, by
vector $\vec{t}$, defined in Fig. 3.5. In this way, the spheres of the **B** layer are
placed in holes of type (a) of layer **A**, shown in Fig. 3.4. The spheres
of the **C** layer are placed over the holes in the **A** layer not occupied by
the spheres from the **B** layer, i.e., of type (b) in Fig. 3.4. The **C** layer is
shifted with respect to the **A** layer by vector $2\vec{t}$, and with respect to
the **B** layer by vector $\vec{t}$, so each sphere of the **C** layer is in contact with
3 spheres of the **B** layer. The spheres of the fourth layer lie directly
above the spheres of the first one.

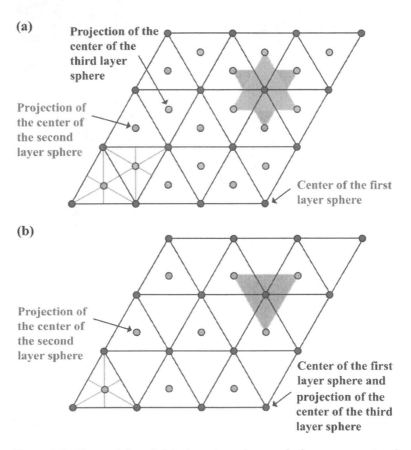

Figure 3.6 Draws (a) and (b) show two close-packed arrangements of equal spheres. The case described in (a) differs from that one in (b) in the positions of spheres of the third layer with respect to the spheres of the first and second layers.

To summarize this part, we can say that in the case shown in Fig. 3.6a we have a *cubic close-packed* (*ccp*) structure that was already introduced as the *fcc* one. This is an **ABCABC**... type structure. Now it is easy to visualize the 12 NNs of an atom in the *fcc* structure; 6 of them belong to the layer in which is placed the atom in consideration, while half of the other 6 belong to the layer below and the other half to the layer above.

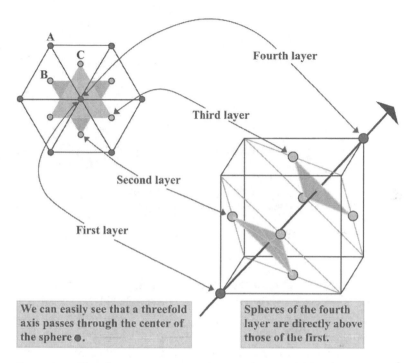

Figure 3.7 The *fcc* structure viewed as a close-packed structure (*ccp*). Three consecutive layers of this structure are marked as **A**, **B**, and **C**.

In the case shown in Fig. 3.6b, we have an *hcp* structure with an **ABAB**... staking sequence of layers. Figure 3.8 shows a part of Fig. 3.6b together with a hexagonal prism. We can see in Fig. 3.8 that the *hcp* structure represents a hexagonal Bravais lattice with two-atom basis. Each atom in this structure has 12 NNs (as is also the case of the *fcc* structure); 6 of them belong to the layer in which is placed the atom in consideration and the other 6 belong to the adjacent layers. The difference between the *ccp* and *hcp* structures consists in the location of the NN atoms that belong to the adjacent layers. In the case of the *ccp* structure three of them occupy holes of type (a) and the other three holes of type (b) (specified in Fig. 3.4) present in the layer to which belongs the atom in consideration. In the case of the *hcp* structure, these 6 NNs occupy holes of type (a), 3 from the top and 3 from the bottom side of the layer. Twelve is the

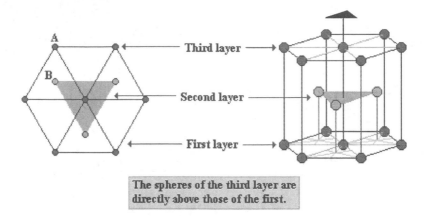

The spheres of the third layer are directly above those of the first.

Figure 3.8 The **ABAB**... stacking sequence of atomic layers in the *hcp* structure.

maximum number of spheres that can be arranged to touch a given sphere. The *hcp* structure will be discussed in more details later.

There are an infinite number of possible ways of close-packing equal spheres, since any sequence of the **A**, **B**, and **C** layers, with no two successive layers alike, represents a possible close-packing arrangement of equal spheres. Therefore, a close-packed structure can be obtained only if two consecutive layers are of a different type. In this case, each sphere touches 12 other spheres and this characteristic of all close-packed structures could be seen already in the case of *fcc* and *hcp* structures. Please note that the only close-packed structure that represents a Bravais lattice with one-atom basis is the *fcc* structure.

Below we will give an example of a close-packed structure, different from the *fcc* and *hcp* structures, which has an **ABACABAC**... layer sequence. This structure is called a *double hexagonal close-packed* (*dhcp*) structure.

3.8 Double Hexagonal Close-Packed Structure

Pearson symbol: *hP4*; **prototype: La.** Two consecutive layers in the *dhcp* structure are of a different type, so it represents indeed one of the close-packed structures with the coordination number

Table 3.4 Lattice parameters of lanthanides that crystallize in the *dhcp* structure under normal conditions. The data for δ-Sm correspond to room temperature and 4.0 GPa

Element	$a(\text{Å})$	$c(\text{Å})$	c/a
α-La	3.7740	12.171	2×1.61
β-Ce	3.681	11.857	2×1.61
α-Pr	3.6721	11.8326	2×1.61
α-Nd	3.6582	11.7966	2×1.61
α-Pm	3.65	11.65	2×1.60
δ-Sm (4.0 GPa)	3.618	11.66	2×1.61

12. Under normal conditions, in the *dhcp* structure crystallize 5 rare earth (RE) metals: lanthanum (α-La), cerium (β-Ce), praseodymium (α-Pr), neodymium (α-Nd), and promethium (α-Pm), all of them lanthanides, and the following actinides: americium (α-Am), curium (α-Cm), berkelium (α-Bk), and californium (α-Cf). Cerium exhibits at room temperature and normal pressure two phases: β and γ (γ-Ce has the *fcc* structure, see Table 3.3). The phase transition from β-Ce to γ-Ce occurs close to the room temperature and β-Ce exists below this temperature. In Table 3.4, we have listed the experimental lattice parameters a and c for La, Ce, Pr, Nd, and Pm, obtained under normal conditions, and for Sm at room temperature and pressure of 4.0 GPa, while Table 3.5 gives the experimental lattice parameters a and c, obtained at ambient conditions, for actinides. The parameters a and c are defined in Fig. 3.9, where we show the hexagonal prism that represents a volume of the *dhcp* structure, which has the same point symmetry as an infinite structure. In this figure, we also show the sequence, **ABACABAC. . .**, of the two-dimensional *hcp* layers.

Table 3.5 Lattice parameters of actinides that crystallize in the *dhcp* structure under normal conditions

Element	$a(\text{Å})$	$c(\text{Å})$	c/a
α-Am	3.468	11.241	2×1.62
α-Cm	3.496	11.331	2×1.62
α-Bk	3.416	11.069	2×1.62
α-Cf	3.390	11.015	2×1.63

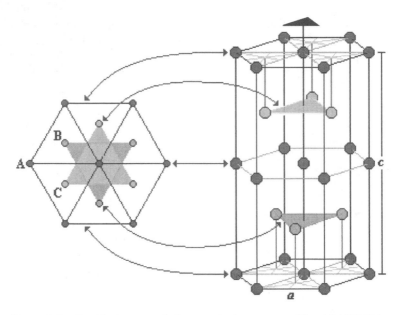

Figure 3.9 Double hexagonal close-packed structure. The **ABACABAC**...
sequence of layers is shown.

In the next section, we will consider the structure of samarium
(α-Sm) at room temperature and pressure.

3.9 Samarium Close-Packed Structure

Pearson symbol: $hR3$, prototype: Sm. Samarium in the α
phase crystallizes in a complex close-packed structure with an
ABABCBCACA... layer sequence. It means that samarium represents the repetition of a unit consisting of 9 two-dimensional *hcp*
layers, as can be seen in Fig. 3.10c. A rhombohedron (inscribed in
the hexagonal prism from Fig. 3.10c) is the smallest unit cell of this
structure, so the α-Sm structure has a rhombohedral space group
symmetry. Its rhombohedral P unit cell contains 3 atoms, while the
triple hexagonal R cell contains 9 atoms. At room temperature and
normal pressure, the cell parameters of the two unit cells, triple
hexagonal and rhombohedral, are $a_h = 3.629$ Å, $c_h = 26.207$ Å and
$a_r = 8.996$ Å, $\alpha_r = 23.22°$, respectively. At high pressures, about half

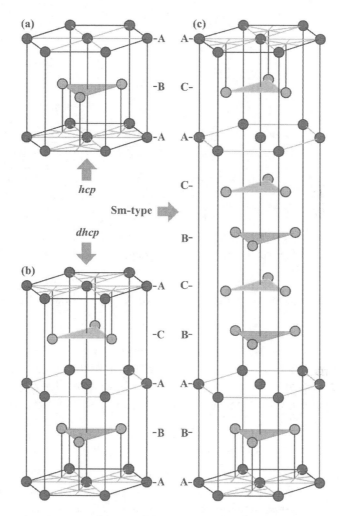

Figure 3.10 Hexagonal prisms for three of the four close-packed structures considered by us: *hcp*, *dhcp*, and Sm-type. In each case, the sequence of the two-dimensional *hcp* layers is shown. The hexagonal prism for the fourth close-packed structure (*ccp*) is displayed in Fig. 3.11.

of RE metals crystallize in the Sm-type structure. They are yttrium (Y), gadolinium (Gd), terbium (Tb), dysprosium (Dy), holmium (Ho), erbium (Er), thulium (Tm), ytterbium (Yb), and lutetium (Lu). In the case of three of them (Y, Tb, and Dy) the following pressure-induced

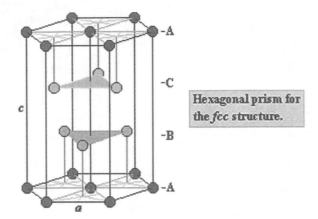

Figure 3.11 Hexagonal prism for the *ccp* (*fcc*) structure (the cell parameters ratio is $c/a = \sqrt{6}$, see Exercise 2.21). The sequence of layers **A**, **B**, and **C** is shown.

sequence of phase transitions is observed at room temperature:

$$hcp \rightarrow \text{Sm-type} \rightarrow dhcp \rightarrow fcc.$$

For yttrium, e.g., the experimentally determined transformations are: from *hcp* to Sm-type at 10–15 GPa, from Sm-type to *dhcp* at 25–28 GPa, and from *dhcp* to *fcc* at 46 GPa. A similar sequence, of pressure-induced phase transitions (at room temperature), is observed for Ho, Er, and Tm:

$$hcp \rightarrow \text{Sm-type} \rightarrow dhcp.$$

It is interesting to mention that the two sequences involve close-packed structures that are, to some degree, mutually related. The *dhcp* and Sm-type structures can be viewed as a certain mixture of the *hcp* and *ccp* (*fcc*) structures.

Figures 3.10 and 3.11 show hexagonal prisms for the four close-packed structures considered by us. The hexagonal prisms for the *hcp*, *dhcp*, and Sm-type structures are drawn in Figs. 3.10a–c, respectively, while in Fig. 3.11 is shown the hexagonal prism for the *fcc* structure. In each case, the sequence of the two-dimensional *hcp* layers is shown. The hexagonal prism displayed in Fig. 3.11 can reproduce the *fcc* structure; however, it has less symmetry (point symmetry) than an infinite *fcc* structure.

3.10 Symmetry Axes

Until now, within all symmetry elements present in three-dimensional crystal structures, we have considered only rotation axes n (n = 1, 2, 3, 4, and 6). To go forward it is necessary to describe other symmetry elements. In a three-dimensional crystal (or in a set of atoms) may be also present reflection (mirror) planes m, rotoinversion axes $\bar{n}$ ($\bar{n}$ = $\bar{1}$, $\bar{2}$, $\bar{3}$, $\bar{4}$, and $\bar{6}$), and rotation axes with center of symmetry, n/m (n/m = $2/m$, $4/m$, or $6/m$). A rotoinversion $\bar{n}$ combines a counterclockwise rotation of an angle of $360°/n$ around an axis with inversion through a point lying on this axis. The symmetry operation $\bar{1}$ is limited to the inversion through a point, which represents the center of symmetry and the rotoinversion axis $\bar{2}$ is identical in its action to the mirror plane that is perpendicular to the axis.

The basic information about rotoinversion axes and rotation axes with center of symmetry is provided in Tables 3.6 and 3.7, respectively. We can observe in Table 3.6 that the inversion point is a center of symmetry for cases when n is odd. The odd rotoinversion axis $\bar{3}$ is present, e.g., in the *sc*, *bcc*, and *fcc* structures. In Fig. 3.7, we have shown a threefold rotation axis of the *fcc* structure. The $\bar{3}$ axis contains this symmetry axis as a sub-element. The $\bar{3}$-inversion point, which is at the same time the center of symmetry, coincides with the geometric center of the cube. In Table 3.7, we can see that the graphical symbol for each rotation axis with center of symmetry, n/m, is a combination of the graphical symbol for the rotation axis n and the graphical symbol for the center of symmetry.

Yet other symmetry elements may be present in an infinite crystal structure. They are glide planes and screw axes. Both, glide reflection through a plane and screw rotation around an axis, involve fractional translations. In what follows, we will describe only those symmetry elements that involve axes.

Screw rotations represent elements of numerous crystallographic space groups. A symmetry operation around an n-fold screw axis, n_p, combines the right-handed rotation of $360°/n$ around this axis with a fractional translation by vector $(p/n)\vec{t}$, where $\vec{t}$, which is pointing in the direction of the screw, is the shortest lattice

Table 3.6 Rotoinversion axes and lower symmetry axes included in them

		Rotoinversion axes, $\bar{n}$	
Printed symbol	Graphical symbol	Lower symmetry axes included in $\bar{n}$	Comments
$\bar{1}$	O		Center of symmetry, inversion center.
$\bar{2}$			Since $\bar{2} = m$, symbol $\bar{2}$ is not used (symbol m, denoting a mirror plane that is perpendicular to the axis, is used instead).
$\bar{3}$	▲	$3, \bar{1}$	
$\bar{4}$	◈ ◪	2	
$\bar{6}$	⬡	$3, \bar{2} = m$	$\bar{6} = 3/m$ The operation $3/m$ combines rotation of $120°$ around the 3 axis with reflection m in a mirror plane that is perpendicular to the axis.

Table 3.7 Rotation axes with center of symmetry and lower symmetry axes included in them

	Rotation axes with center of symmetry, n/m	
Printed symbol	Graphical symbol	Lower symmetry axes included in n/m
$2/m$	⬮	$\bar{1}$
$4/m$	◼ ◆	$\bar{4}, 2, \bar{1}$
$6/m$	⬢	$\bar{6}, \bar{3}, 3, 2, \bar{1}$

translation vector parallel to the axis. Possible values for p are integer numbers between 1 and $n - 1$. In Table 3.8 are listed printed symbols, graphical symbols, and screw vectors for the case of two- and threefold screw axes, while Tables 3.9 and 3.10 give similar information for the four- and sixfold screw axes, respectively. Additionally, in Tables 3.9 and 3.10 is provided the information about the possible lower symmetry axes included in the screw axis.

Table 3.8 Two- and threefold screw axes and the corresponding screw vectors of a right-handed screw rotation

Two- and threefold screw axes, n_p		
Printed symbol	Graphical symbol	Screw vector $(p/n)\vec{t}$
2_1		$(1/2)\vec{t}$
3_1		$(1/3)\vec{t}$
3_2		$(2/3)\vec{t}$

Table 3.9 Fourfold screw axes with the corresponding screw vectors of a right-handed screw rotation and lower symmetry axis included in each case

Fourfold screw axes, 4_p			
Printed symbol	Graphical symbol	Screw vector $(p/4)\vec{t}$	Lower symmetry axes included in 4_p
4_1		$(1/4)\vec{t}$	2_1
4_2		$(1/2)\vec{t}$	2
4_3		$(3/4)\vec{t}$	2_1

We can see in Table 3.8 that the orientation of the "arms" or "blades," ornamenting the triangle that is used as the graphical symbol of the screw axis 3_2, is opposite to the orientation of the "arms" in the case of the screw axis 3_1. This comes from the fact that the right-handed rotation of $120°$ and translation by vector $(2/3)\vec{t}$ is equivalent to a left-handed rotation of $120°$ and translation by vector $(1/3)\vec{t}$. In a similar way, we can justify the orientation of the "arms" in graphical symbols of the screw axes 4_1, 4_3 (Table 3.9) and also 6_1, 6_5 and 6_2, 6_4 (Table 3.10).

When the screw axes 2_1, 4_2, or 6_3 contain centers of symmetry, their printed and graphical symbols are different from those of the noncentrosymmetric case. The basic information about screw axes with center of symmetry is provided in Table 3.11. We can see in this table that the graphical symbol for each screw axis with center of symmetry, n_p/m, combines the graphical symbol for the screw

Table 3.10 Basic information about sixfold screw axes

	Sixfold screw axes, 6_p		
Printed symbol	Graphical symbol	Screw vector $(p/6)\vec{t}$	Lower symmetry axes included in 6_p
6_1		$(1/6)\vec{t}$	$3_1, 2_1$
6_2		$(1/3)\vec{t}$	$3_2, 2$
6_3		$(1/2)\vec{t}$	$3, 2_1$
6_4		$(2/3)\vec{t}$	$3_1, 2$
6_5		$(5/6)\vec{t}$	$3_2, 2_1$

Table 3.11 Basic information about screw axes with center of symmetry

	Screw axes with center of symmetry, n_p/m		
Printed symbol	Graphical symbol	Screw vector $(p/n)\vec{t}$	Lower symmetry axes included in n_p/m
$2_1/m$		$(1/2)\vec{t}$	$\bar{1}$
$4_2/m$		$(1/2)\vec{t}$	$\bar{4}, 2, \bar{1}$
$6_3/m$		$(1/2)\vec{t}$	$\bar{6}, \bar{3}, 3, 2_1, \bar{1}$

axis n_p with the graphical symbol for the center of symmetry. We can also see, that the set of lower symmetry axes included in each case is different from that of the noncentrosymmetric case.

Figure 3.12a shows the presence of a screw axis 3_1 in the infinite crystal of selenium. A right-handed rotation of 120° around the axis is accompanied by a fractional translation by the screw vector $(1/3)\,\vec{t}$, where $\vec{t}$ is defined in the draw. After symmetry operation 3_1, the atoms of coordinates 0 and $\frac{1}{3}$, in terms of basis vector $\vec{c}$, take the position of the atoms of coordinates $\frac{1}{3}$ and $\frac{2}{3}$, respectively (see Fig. 3.12a to the right).

A sixfold screw axis with center of symmetry, $6_3/m$, in the infinite crystal structure of graphite is shown in Fig. 3.12b. The dashed line indicates changes in the location of atoms after a right-handed rotation of 60° around the axis, accompanied by a fractional

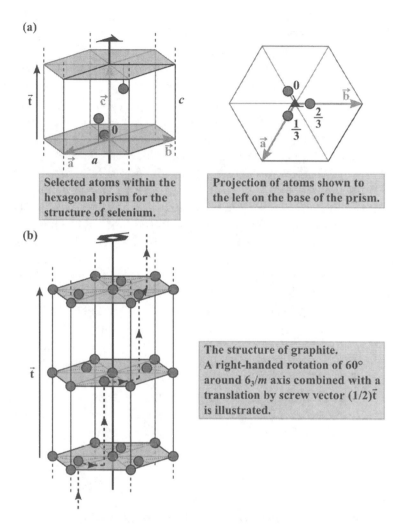

(a)

Selected atoms within the hexagonal prism for the structure of selenium.

Projection of atoms shown to the left on the base of the prism.

(b)

The structure of graphite. A right-handed rotation of 60° around $6_3/m$ axis combined with a translation by screw vector $(1/2)\vec{t}$ is illustrated.

Figure 3.12   (a) Threefold screw axis 3_1 in the infinite selenium structure. The fractions $\frac{1}{3}$ and $\frac{2}{3}$ near the atom projections, in the drawing to the right, are coordinates of these atoms given in terms of the basis vector $\vec{c}$. (b) Sixfold screw axis with center of symmetry, $6_3/m$, in the infinite graphite structure. The dashed line with arrows indicates changes in the positions of atoms after the right-handed rotation of 60° around the axis accompanied by a fractional translation by screw vector $(1/2)\,\vec{t}$. The vector $\vec{t}$, pointing in the direction of the screw, is the shortest lattice translation vector parallel to the axis.

translation by the screw vector $(1/2)\vec{t}$ ($\vec{t}$ is defined in the draw). We can also observe in Fig. 3.12b that the centers of symmetry are equidistant from any pair of adjacent layers.

Some of the symmetry axes present in the *hcp* crystal structure and their relative locations in it are shown in Fig. 3.13. We can see in Fig. 3.13a, for instance, a rotoinversion axis $\bar{6}$ and a threefold rotation axis included in it, while in Fig. 3.13b is shown a sixfold screw axis with center of symmetry, $6_3/m$, and two of the axes included in it, 3 and 2_1. The two axes, $\bar{6}$ and $6_3/m$, are parallel to each other and their relative location is shown in Fig. 3.13c. In the case of the finite volume shown in Fig. 3.13a, the $\bar{6}$-inversion point lies on layer **B**, but in an infinite crystal structure, the $\bar{6}$-inversion points lie on layers **A** and **B**. On the other hand, the centers of symmetry placed on the $6_3/m$ axis are equidistant from layers **A** and **B**. Looking at Table 3.11, we see that the $6_3/m$ axis has a rotoinversion axis $\bar{6}$ included in it. Its inversion points also lie on layers **A** and **B**. So, these inversion points alternate with the centers of symmetry.

Before ending, let us return for a moment to Fig. 3.9 in which we have shown a threefold rotation axis for a set of atoms belonging to the *dhcp* structure. The highest-order symmetry axis of this set of atoms is, however, a rotoinversion axis $\bar{3}$. The $\bar{3}$-inversion point, which is at the same time the center of symmetry, coincides with the geometric center of the hexagonal prism. It is also easy to realize that in the case of an infinite *dhcp* structure, the 3 axis is a sub-element of the $6_3/m$ axis. The centers of symmetry lie on layers **A**.

3.11 Hexagonal Close-Packed Structure

Pearson symbol: *hP*2, prototype: Mg. Now, we will analyze the *hcp* structure in more details. The highest-order symmetry axes of an infinite *hcp* structure are the rotoinversion axis $\bar{6}$ and the screw axis with center of symmetry $6_3/m$ shown in Figs. 3.13a and 3.13b, respectively. A high-symmetry point, $\bar{6}$-inversion point on axis $\bar{6}$, is proposed as the origin of the hexagonal P unit cell (see Fig. 3.14a). An alternative choice of the origin is shown in Fig. 3.14b. In this case, it coincides with a high-symmetry point on axis $6_3/m$ (a center of symmetry). Neither of the two choices for the origin is

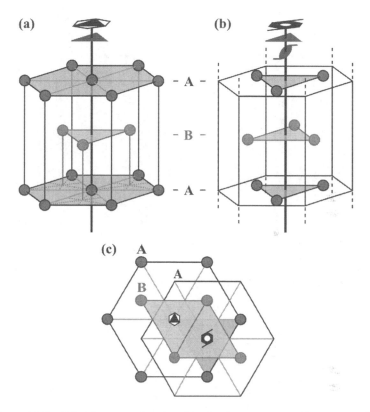

Figure 3.13 (a) A hexagonal prism for the *hcp* structure. Shown are the rotoinversion axis $\bar{6}$ and the threefold rotation axis included in it. The $\bar{6}$-inversion point is located on layer **B** (in an infinite volume the $\bar{6}$-inversion points are located on layers **A** and **B**). (b) Hexagonal prism for the *hcp* structure shifted horizontally with respect to that from (a) in the way shown in (c). The symmetry axis $6_3/m$ and also the lower symmetry axes 3 and 2_1 that are sub-elements of it are shown. The centers of symmetry are equidistant from layers **A** and **B**. (c) Top view of the set of atoms within the hexagonal prism from (a) and the projection of the atoms within the infinite volume specified in (b) on the base of the prism.

considered standard. Interestingly, two alternative choices for origin are proposed in the case of 24 centrosymmetric space groups, while the rest of space groups are described with only one choice of origin. The hexagonal P unit cells (defined by vectors $\vec{a}$, $\vec{b}$, and $\vec{c}$) with the two alternative choices for origin are shown in Figs. 3.14a and 3.14b,

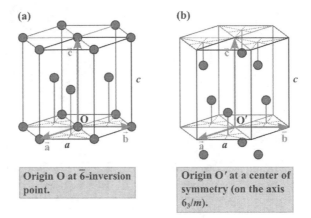

| Origin O at $\bar{6}$-inversion point. | Origin O' at a center of symmetry (on the axis $6_3/m$). |

Figure 3.14 The hexagonal P unit cell of the *hcp* structure defined by the basis vectors $\vec{a}$, $\vec{b}$, and $\vec{c}$. In (a) the origin **O** of the cell coincides with an $\bar{6}$-inversion point, whereas in (b) the origin **O'** coincides with a center of symmetry placed on the $6_3/m$ axis.

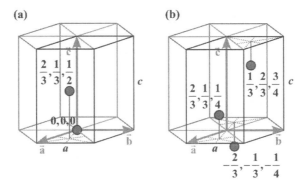

Figure 3.15 Coordinate triplets of the two atoms that are within the hexagonal P unit cell of the *hcp* structure: (a) from Fig. 3.14a and (b) from Fig. 3.14b. The coordinates are expressed in units of a and c.

and the coordinate triplets of the two atoms that are within each cell are given in Figs. 3.15a and 3.15b, respectively.

We will now calculate the c/a ratio in an ideal case when the atoms, considered as hard spheres, touch their NNs. Figure 3.16 shows the hexagonal P unit cell for this case. The three atoms marked as 1, 2, and 3 (from the bottom base) and the three marked

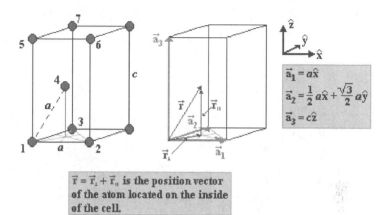

$$\vec{r} = \vec{r}_\perp + \vec{r}_\parallel \text{ is the position vector}$$
of the atom located on the inside
of the cell.

Figure 3.16 Six of twelve NNs (labeled 1–3 and 5–7) of the atom labeled as 4. The 6 NNs are located at the vertices of a hexagonal P unit cell. Vector $\vec{r}$ gives the position of the atom marked as 4.

as 5, 6, and 7 (from the top base) are the NNs of the atom marked as 4, and they are in contact with it. Likewise, atoms marked as 1, 2, and 3 are in contact among themselves. Since the distance between the centers of spheres 1 and 2 is a, then the distance between the centers of spheres 1, 2, or 3 and the center of sphere 4 is also a. To obtain the c/a ratio, we first derive the expression for vectors $\vec{r}_\parallel$ and $\vec{r}_\perp$ that are the components of the position vector $\vec{r}$ of the center of the atom marked as 4 in Fig. 3.16. In Fig. 3.17, we show the plane of the rhombic base of the hexagonal P unit cell which contains $\vec{r}_\perp$. This vector can be expressed as a linear combination of basis vectors $\vec{a}_1$ and $\vec{a}_2$ (see Fig. 3.17). Vectors $\vec{r}_\parallel$ and $\vec{r}_\perp$ can be then written as

$$\begin{cases} \vec{r}_\parallel = \dfrac{1}{2}\vec{a}_3 \\ \vec{r}_\perp = \dfrac{1}{3}(\vec{a}_1 + \vec{a}_2) \end{cases}, \text{ where } \begin{cases} \vec{a}_1 = a\hat{x} \\ \vec{a}_2 = \dfrac{1}{2}a\hat{x} + \dfrac{\sqrt{3}}{2}a\hat{y} \\ \vec{a}_3 = c\hat{z} \end{cases}, \quad (3.6)$$

and can be expressed in terms of orthogonal unit vectors (see Fig. 3.16 to the right) as

$$\begin{cases} \vec{r}_\parallel = \dfrac{1}{2}c\hat{z} \\ \vec{r}_\perp = \dfrac{1}{2}a\hat{x} + \dfrac{\sqrt{3}}{6}a\hat{y} \end{cases}. \quad (3.7)$$

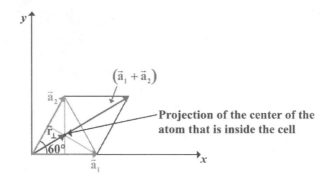

Figure 3.17 The base of the hexagonal P unit cell shown in Fig. 3.16.

Finally, from Eqs. 3.7 we obtain vector $\vec{r}$, which is

$$\vec{r} = \vec{r}_\perp + \vec{r}_\parallel = \frac{1}{2}a\hat{x} + \frac{\sqrt{3}}{6}a\hat{y} + \frac{1}{2}c\hat{z}. \qquad (3.8)$$

Since the module of vector $\vec{r}$ is a (the distance between the centers of atoms 1 and 4), then

$$|\vec{r}| = \sqrt{\frac{a^2}{4} + \frac{3}{36}a^2 + \frac{c^2}{4}} = a \qquad (3.9)$$

and the c/a ratio is

$$\frac{c}{a} = \sqrt{\frac{8}{3}} \cong 1.633. \qquad (3.10)$$

Therefore, in the ideal *hcp* structure the c/a ratio is about 1.63.

Table 3.12 lists lattice parameters a and c for all elements that crystallize in the *hcp* structure at room temperature and normal pressure. As in the case of *bcc* and *fcc* structures all of them are metals. The table reports also lattice parameters of other metals and also helium, obtained at conditions different from normal conditions. We can observe in the table that with the exception of two metals, cadmium and zinc, for the rest of them the c/a ratio is quite close to the ideal value 1.63. The structures of cadmium and zinc are somewhat distorted from the ideal *hcp* structure. The NNs of an atom are not 12 but 6 (the ones from the same layer), while the other 6 atoms, which are placed in adjacent layers, are 10% farther away. However, the symmetry of the *hcp* structure does not depend on the c/a ratio.

Table 3.12 Lattice parameters of metals that crystallize in the *hcp* structure. The data is given at room temperature and normal pressure, unless otherwise specified. Values for helium (^{3}He and ^{4}He) are also included

Element	a (Å)	c (Å)	c/a	Element	a (Å)	c (Å)	c/a
α-Be	2.286	3.585	1.57	Os	2.734	4.320	1.58
Cd	2.979	5.620	1.89	Re	2.761	4.458	1.62
ε-Co	2.507	4.069	1.62	Ru	2.706	4.282	1.58
α-Dy	3.5915	5.6501	1.57	α-Sc	3.3088	5.2680	1.59
α-Er	3.5592	5.5850	1.57	β-Sm (723 K)	3.663	5.845	1.60
α-Gd	3.6336	5.7810	1.59	α'-Tb	3.6055	5.6966	1.58
^{3}He (3.48 K, 0.163 GPa)	3.501	5.721	1.63	Tc	2.738	4.393	1.60
^{4}He (3.95 K, 0.129 GPa)	3.470	5.540	1.60	α-Ti	2.9503	4.6836	1.59
α-Hf	3.1946	5.0511	1.58	α-Tl	3.457	5.525	1.60
α-Ho	3.5778	5.6178	1.57	α-Tm	3.5375	5.5546	1.57
α-Li (78 K)	3.111	5.093	1.64	α-Y	3.6482	5.7318	1.57
α-Lu	3.5052	5.5494	1.58	α-Yb	3.8799	6.3859	1.65
Mg	3.2093	5.2107	1.62	Zn	2.644	4.9494	1.87
α-Na (5 K)	3.767	6.154	1.63	α-Zr	3.2317	5.1476	1.59

It is interesting to note that under normal conditions more than 25% of the elements crystallize in the *hcp* and *dhcp* structures. This information is provided in Table 3.13.

Let us now proceed to calculate the filling factor for the ideal *hcp* structure. Its hexagonal unit cell volume is given by

$$\Omega_0 = (\vec{a}_1 \times \vec{a}_2) \cdot \vec{a}_3 = \begin{vmatrix} \hat{x} & \hat{y} & \hat{z} \\ a & 0 & 0 \\ \frac{1}{2}a & \frac{\sqrt{3}}{2}a & 0 \end{vmatrix} \cdot (c\hat{z}) =$$

$$= \left(\frac{\sqrt{3}}{2}a^2\hat{z}\right) \cdot (c\hat{z}) = \frac{\sqrt{3}}{2}a^2c = \sqrt{2}a^3, \quad \text{where } c = \sqrt{\frac{8}{3}}a \tag{3.11}$$

so

$$(\textit{filling factor})_{hcp} = \frac{2\frac{4}{3}\pi\left(\frac{a}{2}\right)^3}{\sqrt{2}a^3} = \frac{\sqrt{2}}{6}\pi \cong 0.74. \tag{3.12}$$

Note that we have obtained the same result as in the case of the *fcc* (*ccp*) structure. This is the value of the filling factor for any close-

Table 3.13 Crystal structures of all metals that crystallize in dense-packed structures (*fcc, hcp, dhcp,* Sm-type, and *bcc*) under normal conditions. The structures of noble gases at low temperatures are also included

PERIODIC TABLE OF ELEMENTS
Crystal structures

Legend:
- hcp
- fcc
- dhcp
- bcc
- Sm-type
- Mn-type

H																	He
Li β	Be α											B	C	N	O	F	Ne
Na β	Mg											Al α	Si	P	S	Cl	Ar
K	Ca α	Sc α	Ti α	V	Cr α	Mn α	Fe α	Co ε	Ni	Cu	Zn	Ga	Ge	As	Se	Br	Kr
Rb α	Sr α	Y α	Zr α	Nb	Mo	Tc	Ru	Rh	Pd	Ag	Cd	In	Sn	Sb	Te	I	Xe
Cs α	Ba α	La α	Hf α	Ta	W	Re	Os	Ir	Pt	Au	Hg	Tl α	Pb α	Bi	Po	At	Rn
Fr	Ra																

Lanthanides:
La α | γCe / βCe | Pr α | Nd α | Pm α | Sm α | Eu α | Gd α | Tb α' | Dy α | Ho α | Er α | Tm α | βYb / αYb | Lu α

Actinides:
Ac | Th α | Pa | U | Np | Pu | Am α | Cm α | Bk α | Cf α | Es | Fm | Md | No | Lr

packed structure. All of them characterize the maximum number, 12, of the NNs of an atom.

Under normal conditions, more than 40% of elements crystallize in three close-packed structures *fcc*, *hcp*, and *dhcp*, as shown in the periodic table of elements (see Table 3.13).

3.12 Interstices in Close-Packed Structures

We will now examine the *interstices*—empty spaces between atoms (hard spheres)—in close-packed structures. They are of two types, tetrahedral and octahedral. A tetrahedral interstice could be found already in Fig. 3.16. In this figure, the centers of spheres marked as 1, 2, 3, and 4 represent vertices of a tetrahedron. The edges of this tetrahedron are of the same longitude, $2r$ (where r is the sphere radius), so this is a regular tetrahedron. The empty space between the four spheres defining a tetrahedron is what we call a *tetrahedral interstice*. The spheres marked in Fig. 3.16 as 4, 5, 6, and 7 define another regular tetrahedron. The two tetrahedrons have different spatial orientation. The top view of the two types of tetrahedrons is shown in Fig. 3.18a, in which we have the plane of a hexagonal layer **A** and the projection of the centers of spheres of layer **B** on layer **A**. Three vertices of one of the tetrahedrons shown in this figure are found in layer **A** and the fourth in layer **B**. The opposite occurs for the tetrahedron with a second orientation in which three vertices are in layer **B** and the fourth in layer **A**.

The top view of a regular octahedron is visualized in Fig. 3.18b; three of its six vertices are placed in the **A** layer and the other three in the **B** layer. The empty space between the 6 spheres that define the regular octahedron represents the second type of interstices that are present between two different types of layers in a close-packed structure, the so-called *octahedral interstices*. The octahedron edge length is $2r$ like in the case of the tetrahedron edges.

Let us now visualize the interstices present in the two most common close-packed structures. Figure 3.19 shows three of the tetrahedral interstices present in a hexagonal prism for the *hcp* structure. Whereas, in Fig. 3.20, we show two of the tetrahedral and one of the octahedral interstices present in the conventional cubic

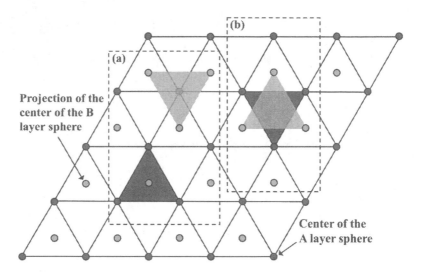

Figure 3.18 Centers of spheres of the **A** layer and projection of the centers of spheres of the **B** layer. (a) Highlighted are the top views of two tetrahedrons with a different spatial orientation, one with three vertices and one vertex projection and the other one with one vertex and three vertex projections. (b) Top view of octahedron bases lying on **A** and **B** layers.

F unit cell of the *fcc* (*ccp*) structure. Additionally, Fig. 3.21 shows the location of all the octahedral interstices present in the cubic *F* unit cell of the *fcc* structure (some of them belong only partially to the cube). There are crystal structures in which some or all of those interstices are filled with atoms. In general, these atoms are of another type than the atoms of the close-packed structure. This gives rise to a large number of compounds that can be described in terms of a close-packing of equal spheres. We will discuss this in details for binary compounds.

So far, we have learned the following about close-packed structures:

(a) Three-dimensional close-packed structures are built of two-dimensional *hcp* layers of equal spheres (see Fig. 3.4). Each sphere of such a layer is in contact with 6 other spheres, which is the maximum possible number of NNs in two dimensions.

(b) The consecutive layers in a three-dimensional close-packed structure are shifted horizontally one with respect to the other.

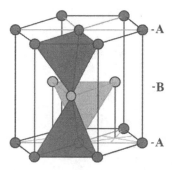

Figure 3.19 Three tetrahedral interstices inside a hexagonal prism for the *hcp* structure. The sequence of layers **A** and **B** is shown.

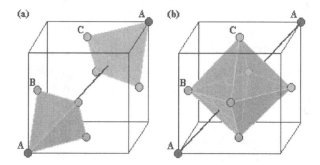

Figure 3.20 Two tetrahedral (a) and one octahedral (b) interstices in a cubic *F* unit cell of the *fcc* (*ccp*) structure. The sequence of layers **A**, **B**, and **C**, which are orthogonal to a body diagonal of the cube, is shown.

As a result, we distinguish three types of layers: **A**, **B**, and **C**, defined in Fig. 3.7.

(c) The spheres of each layer rest in the holes of the layer below. Therefore, each sphere, apart from the 6 NNs in its own layer, has 3 NNs in each of the adjacent layers.

(d) There are tetrahedral and octahedral interstices between two consecutive layers of the close-packed structure.

We know already that metallic elements have the tendency to crystallize in close-packed structures. Moreover, noble gases also crystallize in those structures. The type of bonding between atoms of these elements gives preference for the coordination number 12, as

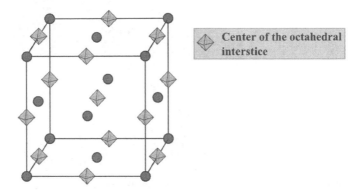

Center of the octahedral interstice

Figure 3.21 Distribution of the octahedral interstices within the cubic *F* unit cell of the *fcc* structure. The interstices that have their centers placed on cube edges belong only partially to the cube.

is indeed achieved in a close-packed structure. We have also learned that various metallic elements crystallize in the *bcc* structure, if not at room temperature, at least at higher temperatures. As we know, each atom in the *bcc* structure has 8 NNs; however, its 6 NNNs are at a distance of only about 15% larger than that to its NNs. Nonmetallic types of bonds—covalent or ionic—require mostly 4 or 6 (less frequently 8) NNs of an atom (ion). The presence of tetrahedral and octahedral interstices in close-packed structures offers the possibility to form bonds with 4 or 6 NNs when an atom (ion) is placed inside a tetrahedral or octahedral interstice, respectively.

Interestingly, a *bcc* crystal structure can be obtained from the *ccp* (*fcc*) structure by placing additional atoms, of the same type as that of the *ccp* structure, in all octahedral and tetrahedral interstices of the *ccp* structure. Figure 3.22a shows a cubic *F* unit cell for the *fcc* (*ccp*) crystal structure. By filling each octahedral and tetrahedral interstice of this cubic cell with the same type of atom as the host one, we obtain the large cube, shown in Fig. 3.22b, composed of 8 small cubes. Each small cube represents a cubic *I* unit cell of the *bcc* structure. Figure 3.23a shows a hexagonal prism for the *fcc* structure depicted in Fig. 3.22a, and Fig. 3.23b shows four hexagonal prisms for the *bcc* structure shown in Fig. 3.22b. These four hexagonal prisms form a large prism, which is related to the large cube from

(a)

(b)

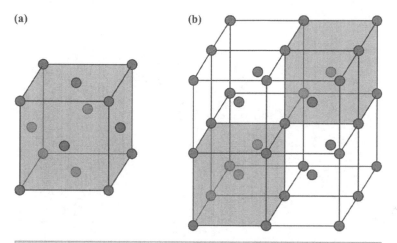

> By placing additional atoms in all tetrahedral and octahedral interstices
> of the *fcc* structure from (a), we obtain the *bcc* structure shown in (b).

Figure 3.22 (a) Cubic F cell of the *fcc* structure. (b) Eight cubic I cells of the *bcc* structure.

Fig. 3.22b in the same way as the hexagonal prism from Fig. 3.23a is related to the cube shown in Fig. 3.22a. The expansion of the *fcc* structure upon filling all its interstices does not change the ratio between the hexagonal cell parameters a and c, therefore

$$(c/a)_{\text{large prism}} = (c/a)_{fcc}. \qquad (3.13)$$

Additionally, from Fig. 3.23b, we can deduce that $a_{bcc} = a_{\text{large prism}}$ and $c_{bcc} = 1/4 c_{\text{large prism}}$, so

$$(c/a)_{bcc} = 1/4 \, (c/a)_{fcc}. \qquad (3.14)$$

Both structures, *bcc* and *fcc*, are of the **ABCABC.** ... type, but, contrary to the case of the *fcc* structure, the *bcc* structure is not close-packed. In general, the atom (or ion) that is located inside the interstice pushes apart the atoms that are at the vertices of the tetrahedron or octahedron. As a consequence, the *hcp* layers (**A**, **B**, and **C**) do not represent any longer close-packed layers but just two-dimensional hexagonal structures. In these layers, the atoms do not touch each other; this occurs in the **A**, **B**, and **C** layers of the *bcc* structure, and also other crystal structures such as *sc*, rhombohedral type, and so on. It is convenient to consider jointly the *fcc, sc, bcc,* and

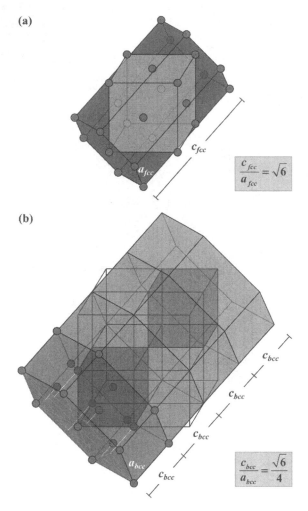

(a)

$$\frac{c_{fcc}}{a_{fcc}} = \sqrt{6}$$

(b)

$$\frac{c_{bcc}}{a_{bcc}} = \frac{\sqrt{6}}{4}$$

Figure 3.23 (a) Hexagonal prism for the *fcc* structure shown in Fig. 3.22a in relation to the cubic *F* unit cell. (b) Four hexagonal prisms for the *bcc* structure shown in Fig. 3.22b. The large prism (of height $4c_{bcc}$) is related to the large cube from Fig. 3.22b.

monatomic rhombohedral type structures structures as they are, in some sense, related. The hexagonal cell parameters a and c of these structures fulfill the following relation

$$(c/a)_{fcc} > (c/a)_{hR1\text{-}Hg} > (c/a)_{sc} > (c/a)_{hR1\text{-}Po} > (c/a)_{bcc}.$$

The largest value of the c/a ratio corresponds to the *fcc* crystal structure in which an atom from each layer has 6 NNs within this layer and 3 in each of the two adjacent layers (12 NNs in total). When the distance between an atom and the atoms closest to it in its own layer is increasing, the hexagonal cell parameter a is also increasing, and the parameter c is decreasing. As a consequence, the c/a ratio is decreasing, and an atom has only 6 NNs placed in the adjacent layers, as is the case of the $hR1$-Hg (α-Hg), $cP1$-Po (sc), and $hR1$-Po (β-Po) structure types. The *bcc* structure has the smallest possible value of the c/a ratio. It has 8 NNs, 6 of them in the adjacent layers, as it was in the previous cases, and the remaining 2 in the next to the adjacent layers in the direction of the basis vector $\vec{c}$, which is orthogonal to the layer planes. Therefore, the layers cannot be closer to each other than they are in the *bcc* structure. The values of the c/a ratio for the *fcc*, *sc*, and *bcc* structures are $\sqrt{6}$, $\sqrt{6}/2$, and $\sqrt{6}/4$, respectively. The c/a ratios for the monatomic rhombohedral type crystal structures $hR1$-Hg and $hR1$-Po fulfill, therefore, the inequality

$$\sqrt{6} > (c/a)_{hR1-\text{Hg}} > \sqrt{6}/2 > (c/a)_{hR1-\text{Po}} > \sqrt{6}/4.$$

In summary, in this segment we have considered two cubic structures, *bcc* and *sc*, which are directly related with the third one, *fcc*. The *bcc* structure may be seen as the *fcc* structure with additional atoms in all its octahedral and tetrahedral interstices, while the *sc* structure may be seen as the *fcc* structure with additional atoms in all its octahedral interstices.

In the next section, we will learn about the diamond structure, which may be seen as the *fcc* structure with 4 additional atoms in half of its 8 tetrahedral interstices. The distribution of the extra atoms guarantees that each atom in the diamond structure has 4 NNs. On the other hand, the diamond structure represents a sequence of layers of the **AABBCCAABBCC**... type, where each layer is a two-dimensional hexagonal (not close-packed) structure. This sequence can be seen as a superposition of two equal sequences of the **ABCABC**... type. It means that we are in presence of a superposition of two *fcc* structures. One of them is translated with respect to the other in such a way that the atoms of each of them occupy half of the tetrahedral interstices of the other. Thus, each atom has 4 NNs, as is required in the case of pure covalent bonding of an element from column IV of the periodic table.

The **A**, **B**, and **C** layers are present also in the case of various important binary compounds. This will be shown in the next chapter on the examples of compounds that crystallize in the zinc blende, wurtzite, NiAs, or NaCl structures. In a crystalline binary compound, the atoms (ions) of a given type form a structure which is at the same time its substructure. In the case of the structures mention above, each of these substructures represents a sequence of layers of the **ABAB**. . . or **ABCABC**. . . type (or only of **AA**. . . type like in the case of cations in NiAs). In general, each layer (**A**, **B**, or **C**) represents a two-dimensional hexagonal structure, which is rarely a true close-packed layer.

In the zinc blende and wurtzite structures, each substructure has an **ABCABC**. . . and **ABAB**. . . staking pattern, respectively, and the atoms of a given type occupy half of the tetrahedral interstices present in the other substructure. In NiAs the substructure of anions has an **ABAB**. . . staking of layers and the cations occupy its octahedral interstices, whereas NaCl is composed of substructures with an **ABCABC**. . . layer stacking. The ions from one substructure of NaCl occupy the octahedral interstices present in the other substructure. All these structures will be described in details in the next chapter.

3.13 Diamond Structure

Pearson symbol: *cF*8, prototype: C. Four elements, from column IV of the periodic table, crystallize in the diamond structure, namely carbon, silicon, germanium, and gray tin (α-Se, which is one of the two allotropes of tin at normal pressure and temperature). The atoms of each of these elements have four electrons in the outermost shell (the so-called valence shell). By completing this shell with four additional electrons those atoms can achieve a state of the highest stability. This stability is reached in the crystal of each of these elements in which an atom is surrounded by four neighboring atoms that in turn form covalent bonds (represented schematically in Fig. 3.24) with it. In the diamond structure, each atom shares four electrons with its 4 NNs and each of these neighbors shares an electron with the atom under consideration. Therefore, all atoms

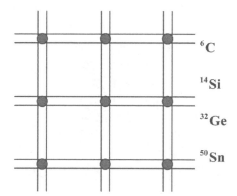

^{6}C

^{14}Si

^{32}Ge

^{50}Sn

Figure 3.24 Two-dimensional schematic diagram of covalent bonds in the diamond structure.

can complete the 4 electrons that were lacking to achieve the highest stability.

The neighborhood of an atom in the diamond structure is shown in Fig. 3.25. The four NNs of each atom of an element that crystallizes in this structure are placed at the vertices of a regular tetrahedron that has the atom under consideration in the center, as shown in Fig. 3.25a. The regular tetrahedron is easier to draw if we place it inside a cube, as was done in Fig. 3.25b. In addition, Fig. 3.25c shows a three-dimensional schematic representation of covalent bonds between an atom and its 4 NNs in the diamond structure.

We should observe that the cubic volume that we have drawn in Fig. 3.25b does not, of course, represent a unit cell of the diamond structure, since it does not have atoms in all its vertices. However, we can easily locate this volume within the cubic unit cell of this structure. Figure 3.26 shows two of the 4 possible positions of the small cube inside the diamond cubic unit cell. We may also observe in Fig. 3.26 that this unit cell is just the cubic F unit cell like in the case of the *fcc* structure, but with 4 additional atoms placed inside (on the body diagonals). The distance between each additional atom and its nearest cube vertex is 1/4 of the cube body diagonal, and those additional atoms occupy 4 of the 8 tetrahedral interstices present in the cubic F unit cell of the *fcc* structure. The tetrahedral interstices of the *fcc* structure have been already considered by us in the previous section. In that opportunity, the *fcc* structure was seen

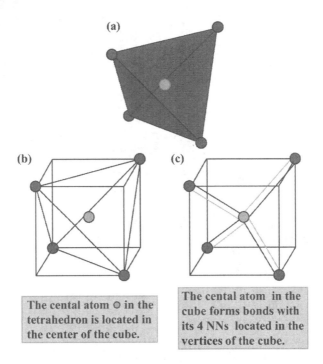

The cental atom ⊙ in the tetrahedron is located in the center of the cube.

The cental atom in the cube forms bonds with its 4 NNs located in the vertices of the cube.

Figure 3.25 (a) A tetrahedron defined by the NNs of an atom in the diamond structure. (b) The tetrahedron from (a) inscribed in a cube. (c) Three-dimensional schematic representation of covalent bonds between an atom and its 4 NNs.

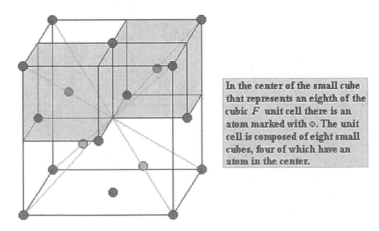

In the center of the small cube that represents an eighth of the cubic F unit cell there is an atom marked with ⊙. The unit cell is composed of eight small cubes, four of which have an atom in the center.

Figure 3.26 Two small cubes from Fig. 3.25b placed in two of the four possible positions inside a cubic F unit cell of the diamond structure.

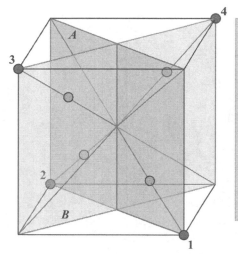

The 4 body diagonals of the cube (which represents the cubic *F* unit cell of the diamond structure) define two vertical planes as shown in the figure. On the plane *A* there are two ○ atoms. These atoms are located on the diagonals, at a distance equal to 1/4 of their longitudes from vertices 1, 2 of the bottom base of the cube. Likewise, two atoms are located on plane *B*, but now equally distant from vertices 3, 4 of the upper base.

Figure 3.27 Relative positions of atoms belonging to the diamond structure. The 4 atoms that are inside the cubic *F* unit cell are distributed in two vertical planes defined by the body diagonals of the cube.

as a sequence of two-dimensional *hcp* layers with an **ABCABC**... staking pattern. In Fig. 3.20a, we have shown two examples of tetrahedral interstices present in the cubic *F* cell of the *fcc* structure.

To help visualize the positions of the 4 atoms that are inside of the cubic *F* unit cell of the diamond structure, we have drawn in Fig. 3.27 two mutually orthogonal vertical planes *A* and *B*. Each plane is defined by two body diagonals of the cube and the 4 atoms are placed on these diagonals in the way explained in this figure.

It is obvious that the neighborhood of each atom in the diamond structure is the same. This can be verified by drawing two cubic *F* unit cells, I and II, of the diamond structure in such a way that cube II is shifted with respect to cube I along one of its body diagonals, to a segment equal in length to 1/4 of the diagonal length. This is partially shown in Fig. 3.28a, where the cube I and the cross section of the cube II in a plane defined by its body diagonals are drawn. We can see in this figure that the atoms that are located on the body diagonals of the cubes have the same spatial distribution as the atoms from vertices and faces. Indeed, one of the atoms that are in the interior of cube I (see Fig. 3.28a) coincides with

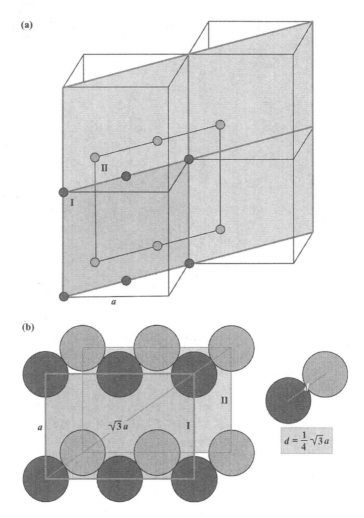

Figure 3.28 (a) Comparison of the distribution of atoms from vertices and faces of the cubic F unit cells with those from their body diagonals in the diamond structure. (b) Cross sections of atoms (considered as hard spheres) from (a) are shown. In this figure, the equivalency between the relative distributions of atoms of each type (those from vertices and faces of the cube and those from its diagonals) is visualized.

an atom located at a vertex of cube II and the other one with the atom located at a face, and *vice versa*. Thus, all atoms in the diamond structure have the same neighborhood. This is additionally illustrated in Fig. 3.28b, where we show the cross sections of atoms from Fig. 3.28a (considered as hard spheres) in a plane defined by the body diagonals of the cubes. Finally, we can say that this arrangement of atoms in the diamond structure allows each atom to be in the center of a regular tetrahedron with 4 NNs (located at the vertices of the tetrahedron) covalently bonded to it.

Summarizing the above, we can say that the diamond structure is just a superposition of two *fcc* substructures that are shifted one with respect to the other in the way described above. Each sub-structure may be seen as a sequence of layers with an **ABCABC**... staking pattern. The two substructures are shifted one with respect to the other, to a segment shorter than the distance between two consecutive layers in the substructures, in the direction orthogonal to the layer planes. Thus, the diamond structure represents indeed a sequence of layers of the **AABBCCAABBCC**... type, as is easy to see in Fig. 3.29, where we have added the cross sections of the **A**, **B**, and **C** layer planes to Fig. 3.28b.

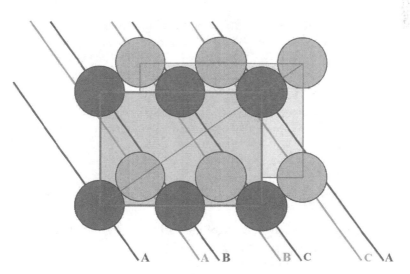

Figure 3.29 Left part of Fig. 3.28b with the cross sections of the **A**, **B**, and **C** layer planes added to it. In this figure, the sequence of layers is easily seen.

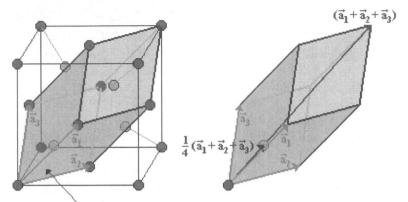

This is a rhombohedral unit cell and it has 2 atoms.

Figure 3.30 Cubic F and rhombohedral P unit cells of the diamond structure (left). In the figure, is also shown the location of the two atoms belonging to the rhombohedral P unit cell (right).

The smallest unit cell of the diamond structure is, of course, the primitive rhombohedral unit cell, like in the case of the *fcc* structure, but in this case, the cell contains 2 atoms, as shown in Fig. 3.30. Therefore, the diamond structure can be seen as a *cF* Bravais lattice with two-atom basis.

Finally, let us specify the location of atoms belonging to the diamond cubic F unit cell. Figure 3.31a shows the coordinate triplets of the 8 atoms that are within this unit cell, and Fig. 3.31b shows the projection of these atoms on the cell base. The coordinates of the 8 atoms, in Fig. 3.31a, are given in terms of the cubic basis vectors $\vec{a}$, $\vec{b}$, and $\vec{c}$, and the fractions near the atom projections in Fig. 3.31b represent the coordinates of these atoms expressed in terms of the basis vector $\vec{c}$.

3.14 Arsenic Structure

Pearson symbol: *hR*2, prototype: As. At normal conditions, arsenic crystallizes in the *hR*2-As (α-As) structure, which has two atoms in the rhombohedral P unit cell. The information about its lattice parameters and atomic basis is given in Table 2.5 (see Section 2.11.4). Figure 3.32a shows the basis vectors $\vec{a}_1$, $\vec{a}_2$, and $\vec{a}_3$

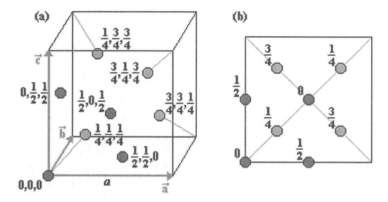

Figure 3.31 (a) Coordinate triplets of the eight atoms within the cubic F unit cell of the diamond structure. The coordinates are expressed in units of a. (b) Projection of atoms on the cell base. The fraction nearby the projection of an atom represents its coordinate in units of $c = a$.

which define the rhombohedral P unit cell of the rhombohedral Bravais lattice hR for this structure. The parameters a and c defined in this figure can be deduced from Table 2.5 and are 3.762 Å and 10.542 Å, respectively. It is easy to realize that the draw from Fig. 3.32c is obtained from Fig. 3.32a by attaching to each lattice point the 2-atom basis shown to the right of Fig. 3.32a, in the way that the center of one of the basis atoms overlaps a lattice point. Note that we have assumed here a non-conventional origin for the unit cell. The conventional origin should coincide with a center of symmetry of the structure, which lies between (is equidistant to) layers **B** and **C** shown in Fig. 3.32c.

At room temperature and pressure of about 25 GPa, arsenic undergoes a phase transition to the monatomic sc structure known as the $cP1$-Po (α-Po) structure. In order to describe this structure, we propose a fcc lattice instead of the sc lattice (see Section 2.11.3); then the atomic basis is composed of 2 atoms. Figure 3.32b shows the hexagonal prism for this fcc lattice and the basis vectors $\vec{a}'_1$, $\vec{a}'_2$, and $\vec{a}'_3$, which define its rhombohedral P unit cell. The experimental values for the parameters a' and c', defined in Fig. 3.32b, are 3.606 Å and 8.833 Å, respectively. They are, respectively, 4% and 16% smaller than the corresponding parameters a and c at normal conditions. By attaching to each fcc lattice point a 2-atom basis

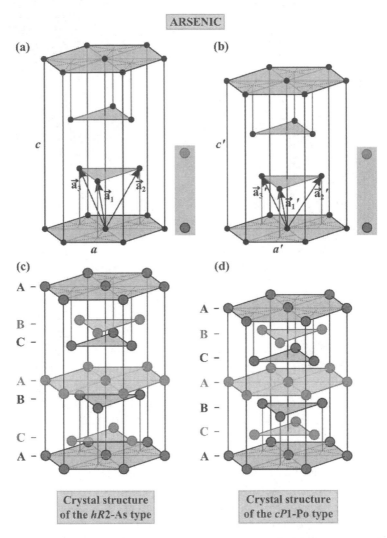

ARSENIC

(a)

c

$\vec{a}_3$ $\vec{a}_1$ $\vec{a}_2$

a

(b)

c'

$\vec{a}_3{}'$ $\vec{a}_1{}'$ $\vec{a}_2{}'$

a'

(c)

A –
B –
C –
A –
B –
C –
A –

Crystal structure
of the *hR*2-As type

(d)

A –
B –
C –
A –
B –
C –
A –

Crystal structure
of the *cP*1-Po type

Figure 3.32 (a) Hexagonal prism for the rhombohedral Bravais lattice *hR* of the normal phase of arsenic. To the right, the atomic basis is shown. (b) Hexagonal prism for the *fcc* lattice proposed to describe the *sc* structure of arsenic at room temperature and pressure of 25 GPa. The atomic basis is shown to the right. (c) Crystal structure of arsenic at normal conditions. (d) Crystal structure of arsenic at room temperature and pressure of 25 GPa. Both figures to the right were rescaled to keep $a' = a$.

(see Fig. 3.32b), the draw from Fig. 3.32d is obtained, which shows two hexagonal prisms for the *sc* structure, one on top of the other.

In both phases of arsenic, normal and high pressure, its structure represents a sequence of layers of the **ACBACB**... type. Each layer consists of a two-dimensional, not close-packed, hexagonal structure. Contrary to the normal phase (Fig. 3.32c), in the case of the high pressure *sc* structure (Fig. 3.32d) all layers are equidistant.

3.15 Selenium Structure

Pearson symbol: *hP3*, prototype: Se. At normal conditions, selenium crystallizes in the *hP3*-Se (γ-Se) structure, which has three atoms in the hexagonal *P* unit cell. Figure 3.33a shows the projection of the three atoms on the cell base (the fraction nearby each projection of an atom represents its coordinate in units of *c*). The conventional origin **O** is placed on the screw axis 3_1 (see Fig. 3.12a). In Fig. 3.33b (top), we show the projection of two hexagonal prisms, for the normal phase of selenium, on the prism bases. One of those prisms contains the hexagonal *P* unit cell with the conventional origin **O** and the other one contains the unit cell with a non-conventional origin **O**′, which coincides with the center of an atom located on the cell base. The basis vectors $\vec{a}$, $\vec{b}$, and $\vec{c}$, shown in Fig. 3.33b (bottom), define the hexagonal *P* unit cell with origin **O**′. The projections of atoms of coordinates $\frac{1}{3}$ and $\frac{2}{3}$ along the $\vec{c}$ axis in such a cell lie on the heights of the equilateral triangles shown in Fig. 3.33b (top). Note, that their location is different from that in Fig. 3.33a. The experimental lattice parameters *a* and *c* [defined in Fig. 3.33b (bottom)] for the normal phase of selenium are 4.368 Å and 4.958 Å, respectively.

At room temperature and high pressures (above 42 GPa) selenium has the *hR1*-Po (β-Po) structure. The information provided in Fig. 3.33c is similar to that given in Fig. 3.33b, but now the projections of atoms of coordinates $\frac{1}{3}$ and $\frac{2}{3}$ along the $\vec{c}$ axis coincide with the geometric centers of the equilateral triangles shown in Fig. 3.33c (top), which is the case of the crystal structure with a rhombohedral space-group symmetry. The lattice parameters *a*′ and *c*′ at 46 GPa are 4.658 Å and 3.858 Å, respectively ($c'/a' = 1.207$ is less

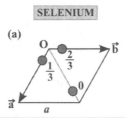

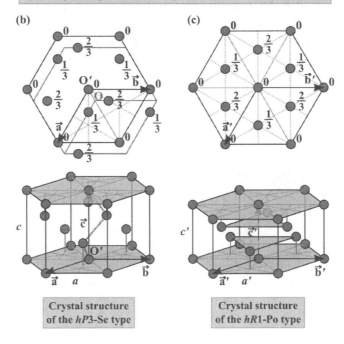

At normal conditions the positions of the three Se atoms (within the hexagonal *P* unit cell) in units of *a* and *c* are: 0.2254, 0, 0.3333; 0, 0.2254, 0.6667; 0.7746, 0.7746, 0.

Figure 3.33 (a) Base of the hexagonal *P* unit cell with a conventional origin O for the structure of selenium at normal conditions. The projection of the three atoms within the unit cell on its base is shown. The fraction nearby the projection of each atom represents its fractional coordinate along the $\vec{c}$ axis. (b) Structure of selenium at normal conditions. In the upper part is shown the projection, on the prism base, of atoms from the hexagonal prism shown in the lower part of the figure. The second hexagon corresponds to the projection of a hexagonal prism (not shown) shifted with respect to that shown in the figure. (c) Structure of selenium at room temperature and pressure of 46 GPa. In the upper part is shown the projection, on the prism base, of atoms located in the hexagonal prism shown in the lower part of the figure.

than $\sqrt{6}/2 = 1.225$, see Section 3.12). The lattice parameter a' increases about 7% and c' decreases about 22% in comparison to their respective values a and c for the normal phase of selenium.

Let us do a final comment about the NNs of an atom in both, considered in this section, phases of selenium. We can see in Fig. 3.33b (top) that the projection (on the prism base with O' in the center) of the centers of atoms, of coordinates $\frac{1}{3}$ along the $\vec{c}$ axis, define an equilateral triangle of edge a. A similar equilateral triangle, of edge a', may be defined in Fig. 3.33c (top) by the centers of atom projections labeled as $\frac{1}{3}$. The fundamental difference between the two situations is such that only in the second case the geometric center of the equilateral triangle coincides with the center of a selenium atom. As a consequence, in the case of selenium at room temperature and pressure of 46 GPa, there are 6 atoms of coordinates $\pm\frac{1}{3}$ along the $\vec{c}$ axis, which are at the shortest distance from the atom placed at the origin of the unit cell defined in Fig. 3.33c (bottom). For the normal phase of selenium, the NNs of the atom placed at the origin O' [see Fig. 3.33b (bottom)] are the 2 closest atoms of coordinates $\pm\frac{1}{3}$ along the $\vec{c}$ axis, therefore each selenium atom has only 2 NNs.

3.16 Graphite Structure

Pearson symbol: $hP4$, prototype: C. Graphite is a crystal composed of carbon honeycomb layers (*graphene layers*) stacked one on top of the other in an **ABABAB**... type configuration (*Bernal configuration*). Layer **A** is shifted with respect to layer **B**, as shown in Fig. 3.34a. The space group of the graphite structure is the same as that of *hcp* and *dhcp* structures. Its Hermann–Mauguin short symbol is $P6_3/mmc$. As an example of a Hermann–Mauguin symbol for the three-dimensional case, the symbol $P6_3/mmc$ will be described below in more details. Figures 3.34b and 3.34c show, respectively, the location of the $\bar{6}$ and $6_3/m$ axes in graphite. A similar draw for the *hcp* structure was shown in Figs. 3.13a and 3.13b. As it was there, also now two alternative origins of the unit cell are proposed for the graphite structure. One of them coincides with an $\bar{6}$-inversion point and the other one with a center of symmetry placed on the $6_3/m$ axis

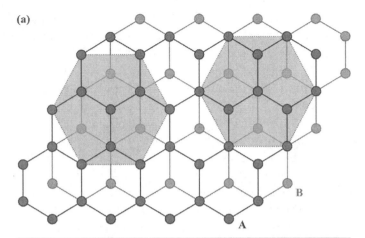

(a)

B

A

Top view of graphite. In the figure it is shown how the honeycomb layer B is shifted with respect to the honeycomb layer A.

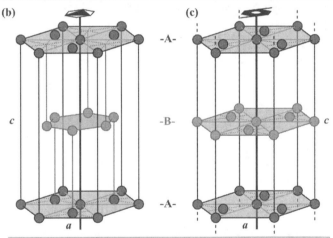

(b)

-A-

c

-B-

-A-

a

(c)

-A-

-B-

c

-A-

a

At normal conditions carbon crystallizes in the $hP4$-C structure, in which the honeycomb layers A and B are distant by 3.3545 Å.

Figure 3.34 (a) Relative position of the projections of the honeycomb **A**, **B** layers in the graphite structure. (b) Hexagonal prism for the graphite structure. The rotoinversion axis $\bar{6}$ is shown. The $\bar{6}$-inversion point is located on layer **B** (in an infinite volume, the $\bar{6}$-inversion points are located on layers **A** and **B**). (c) Hexagonal prism shifted horizontally with respect to that from (b) in the way shown in (a). The symmetry axis $6_3/m$ is shown. The centers of symmetry are equidistant from layers **A** and **B**.

Table 3.14 Sets of symmetry directions in a three-dimensional hexagonal lattice corresponding to a given position in the Hermann–Mauguin space-group symbol

Primary	Secondary	Tertiary
	Position of symmetry direction in the Hermann–Mauguin symbol for a three-dimensional hexagonal lattice	
[001]	$\left\{\begin{array}{c}[100]\\ [010]\\ [\bar{1}\bar{1}0]\end{array}\right\}$	$\left\{\begin{array}{c}[1\bar{1}0]\\ [120]\\ [\bar{2}\bar{1}0]\end{array}\right\}$

(note that this was shown already in Fig. 3.14 for the *hcp* structure). In graphite, the $\bar{6}$-inversion points lie on the honeycomb **A**, **B** layers and the centers of symmetry are equidistant from those layers.

Let us now describe the full $P6_3/m\,2/m\,2/c$ and short $P6_3/mmc$ Hermann–Mauguin space-group symbols. The symmetry directions in a three-dimensional hexagonal Bravais lattice, corresponding to symmetry elements (axes and planes) that are included in different positions (primary, secondary, and tertiary) of the Hermann–Mauguin symbol, are listed in Table 3.14. In the case of planes, the lattice directions are parallel to their normals. The symmetry direction [001] in the *hcp* and *dhcp* structures is orthogonal to the planes of **A** and **B** layers (two-dimensional close-packed structures) and in the case of graphite, it is orthogonal to the honeycomb planes. From Table 3.14, we can read that the secondary and tertiary lattice symmetry directions lie on planes orthogonal to the [001] direction. They coincide with the lattice directions specified in Table 1.2 for a two-dimensional hexagonal lattice. Let us now return to the description of the $P6_3/m2/m2/c$ symbol. In the primary position of the symbol, we have the 6_3 axis and mirror plane *m* orthogonal to it. The two symmetry symbols are separated by a slash which signalizes that the symmetry axis and the normal to the symmetry plane are mutually parallel. In the secondary position of the symbol, it is revealed the presence of twofold rotation axes in the equivalent directions specified in the penultimate column of Table 3.14, and mirror planes *m* orthogonal to those directions. Finally, from the tertiary position of the symbol, we learn that there are twofold rotation axes in the directions specified in the last column of Table 3.14, and glide planes c orthogonal to those directions (a glide reflection through a plane c is a combination of a mirror reflection through a plane containing a $\vec{c}$ basis vector with the translation by vector $1/2\vec{c}$). In the short Hermann–Mauguin symbol $P6_3/mmc$ the symmetry information of the full symbol is reduced.

3.17 Atomic Radius

We have shown all along this chapter, on the example of the most important crystal structures for elements, how the atomic radius can be determined. It is not the purpose of this book to go deeper into the subject of atomic radii; therefore we present only a very shortened and simplified introduction to the subject. As we remember, the atom in a crystal is considered as a hard, impenetrable sphere and in the case of an element its radius is given by half of the distance between NNs, which is determined by the experimentally obtained lattice parameters. In this chapter, we have considered metals, noble gases, and the elements from column IV (and also some elements from columns V and VI) of the periodic table. In each of those cases, the type of bonding is different and the coordination number is related to the type of bonding. We could learn that the metallic bonding prefers the coordination number 12. Some metals crystallize also in the *bcc* structure with coordination number 8. The elements from column IV characterize pure covalent bonding, and in the case of them, each atom requires usually 4 NNs (graphite represents an exception—each of its atoms forms covalent bonds with 3 NNs located in the same honeycomb layer). In the case of covalent bonding in crystals of arsenic and selenium, an atom requires 3 and 2 NNs, respectively. On the other hand, the predominantly ionic and partially ionic and partially covalent bonds appear in the case of compounds and will be discussed in Chapter 4 on the example of binary compounds. The radius of an atom (ion) in a compound depends strongly on its oxidation state. In summary, the radius of an atom in a crystal depends on its oxidation state, the type of bonding between atoms, and the coordination number.

In Table 3.15, we list experimental lattice constants, NN inter-atomic distances, and covalent radii (all parameters obtained under

Table 3.15 Lattice constants of elements that crystallize in the diamond structure under normal conditions. In addition, the NN distances, d, and the covalent radii, r_{cov}, are given

Element	a (Å)	d (Å)	r_{cov} (Å)
C	3.566986	1.545	0.772
α-Si	5.430941	2.352	1.176
α-Ge	5.657820	2.450	1.225
α-Sn	6.4892	2.810	1.405

normal conditions) of elements that crystallize in the diamond structure. The covalent radius for each element is calculated as half of the distance d between NNs determined by the experimental lattice constant a according to the expression

$$d = \frac{1}{4}\sqrt{3}a. \tag{3.15}$$

In a similar way, we have calculated the metallic radii for all metals that crystallize in structures with coordination number 12. Since the system of metallic radii is set up for the coordination number 12, for those metals that crystallize in the *bcc* structure, and therefore, have coordination number 8, we have made a correction (commonly used by chemists) consisting in increasing their radii by 3%.

The NN interatomic distances and metallic radii of all metals considered in this chapter are listed in Tables 3.16 and 3.17, respectively. We can observe in Table 3.16 that in the case of metals that crystallize in the *hcp*, *dhcp*, and Sm-type structures, we are giving two values for the interatomic distances. The upper value corresponds to the distance of an atom to its 6 NNs located in the same layer to which the atom belongs, and the lower value corresponds to the distance to its 6 NNs from the adjacent layers. We can see in this table that only in the case of cadmium and zinc, the two values are substantially different (by about 10%). The metallic radii reported in Table 3.17, for metals that crystallize in the *hcp* structure, were calculated using the average value for the NN distance.

Table 3.16 Nearest neighbor interatomic distances (in Angstroms) of metals that crystallize in dense-packed structures under normal conditions. The values have been calculated using the data from Tables 3.2–3.5 and 3.12. In the case of metals that crystallize in the *hcp*, *dhcp*, and Sm-type structures, we report two interatomic distances: the distance to the 6 NNs from the same layers and the distance to the 6 NNs from adjacent layers

PERIODIC TABLE OF ELEMENTS
Nearest neighbor interatomic distances in Angstroms

1	2	3	4	5	6	7	8	9	10	11	12	13	14	15	16	17	18
H																	He
Li 3.039	Be 2.286 / 2.226											B	C	N	O	F	Ne
Na 3.716	Mg 3.209 / 3.197											Al 2.863	Si	P	S	Cl	Ar
K 4.608	Ca 3.952	Sc 3.309 / 3.254	Ti 2.950 / 2.896	V 2.619	Cr 2.498	Mn (2.244–2.911)	Fe 2.482	Co 2.507 / 2.497	Ni 2.492	Cu 2.556	Zn 2.908 / 2.644	Ga	Ge	As	Se	Br	Kr
Rb 4.941	Sr 4.302	Y 3.648 / 3.557	Zr 3.232 / 3.179	Nb 2.858	Mo 2.725	Tc 2.738 / 2.706	Ru 2.706 / 2.650	Rh 2.689	Pd 2.751	Ag 2.889	Cd 2.979 / 3.295	In	Sn	Sb	Te	I	Xe
Cs 5.318	Ba 4.350	La 3.774 / 3.742	Hf 3.195 / 3.127	Ta 3.861	W 2.741	Re 2.761 / 2.740	Os 2.734 / 2.675	Ir 2.715	Pt 2.775	Au 2.884	Hg	Tl 3.457 / 3.408	Pb 3.500	Bi	Po	At	Rn
Fr	Ra 4.458	Ac 3.755															

Ce 3.649	Pr 3.672 / 3.639	Nd 3.658 / 3.627	Pm 3.65 / 3.59	Sm 3.629 / 3.587	Eu 3.969	Gd 3.634 / 3.572	Tb 3.606 / 3.528	Dy 3.592 / 3.504	Ho 3.578 / 3.504	Er 3.559 / 3.467	Tm 3.538 / 3.447	Yb 3.878	Lu 3.505 / 3.434
Th 3.595	Pa	U	Np	Pu	Am 3.468 / 3.451	Cm 3.496 / 3.478	Bk 3.416 / 3.398	Cf 3.390 / 3.378	Es 4.066	Fm	Md	No	Lr

Table 3.17 Metallic radii (in Angstroms) of metallic elements that under normal conditions crystallize in dense-packed structures (*fcc, hcp, dhcp,* Sm-type, and *bcc*). In the case of elements that crystallize in the *hcp, dhcp,* and Sm-type structures, the radius is half of the average value for the NN interatomic distance

PERIODIC TABLE OF ELEMENTS
Metallic radii in Angstroms

1	2	3	4	5	6	7	8	9	10	11	12	13	14	15	16	17	18
H																	He
Li 1.565	Be 1.128											B	C	N	O	F	Ne
Na 1.914	Mg 1.602											Al 1.432	Si	P	S	Cl	Ar
K 2.373	Ca 1.976	Sc 1.641	Ti 1.462	V 1.349	Cr 1.287	Mn 1.26	Fe 1.278	Co 1.251	Ni 1.246	Cu 1.278	Zn 1.388	Ga	Ge	As	Se	Br	Kr
Rb 2.544	Sr 2.151	Y 1.801	Zr 1.603	Nb 1.472	Mo 1.404	Tc 1.361	Ru 1.339	Rh 1.345	Pd 1.375	Ag 1.445	Cd 1.568	In	Sn	Sb	Te	I	Xe
Cs 2.739	Ba 2.240	La 1.879	Hf 1.580	Ta 1.473	W 1.412	Re 1.375	Os 1.352	Ir 1.357	Pt 1.387	Au 1.442	Hg	Tl 1.716	Pb 1.750	Bi	Po	At	Rn
Fr	Ra 2.296	Ac 1.878															

La 1.879	Ce 1.825	Pr 1.828	Nd 1.821	Pm 1.81	Sm 1.804	Eu 2.044	Gd 1.801	Tb 1.783	Dy 1.774	Ho 1.766	Er 1.757	Tm 1.746	Yb 1.939	Lu 1.735
Ac 1.878	Th 1.797	Pa	U	Np	Pu	Pm 1.81	Am 1.730	Cm 1.744	Bk 1.704	Cf 1.692	Es 2.033	Fm	Md	No	Lr

Problems

Exercise 3.1 Calculate the filling factor for the diamond structure.

(a) Draw a cross section of the cubic F unit cell for the diamond structure which contains the points of contact between the atoms considered as hard spheres.

(b) Express the covalent atomic radius of the atom as a function of the lattice constant a and calculate the filling factor for the diamond structure.

(c) Make a comparison between the filling factor for the diamond structure and the filling factors for the *fcc* (or ideal *hcp*), *bcc*, and *sc* structures. What is the coordination number in each case?

Exercise 3.2 Figure 2.3a (Section 2.2) shows a regular tetrahedron inscribed in a cube and three twofold rotation axes present in it. Which symmetry axes of the tetrahedron do contain the twofold rotation axes as sub-elements?

Exercise 3.3 Figures 3.35a and 3.35b show hexagonal prisms for a *dhcp* structure. One of them is shifted with respect to the other by 1/4 of the prism heights.

(a) What is the highest-order symmetry axis of the set of atoms shown in Fig. 3.35a? Place the graphical symbol of such an axis in the draw.

(b) Repeat (a) for the case shown in Fig. 3.35b.

Exercise 3.4 The threefold rotation axis shown in Fig. 3.9 for the *dhcp* crystal structure represents a lower symmetry axis included in the $6_3/m$ symmetry axis of the infinite *dhcp* structure. In a draw, similar to that from Fig. 3.12b for the infinite graphite structure, show the changes of positions of selected atoms from layers **B** and **C** after a right-handed screw rotation of $60°$ around the $6_3/m$ axis. Additionally, specify the location of the centers of symmetry.

Exercise 3.5 In the draw from Fig. 3.13b, for an infinite *hcp* structure, do a similar demonstration to that from Fig. 3.12b, but now for the screw axis 2_1.

Exercise 3.6 Inside the hexagonal prism for the *dhcp* structure shown in Fig. 3.35a:

(a) Draw the hexagonal P unit cell and the basis vectors $\vec{a}$, $\vec{b}$, and $\vec{c}$ defining it. What is the highest-order symmetry axis of the infinite *dhcp* structure that passes through the origin of the cell that you have drawn? Which sub-element of this axis is included in the site symmetry symbol of the origin? Is the origin of this cell a conventional origin?

(b) Place the coordinate triplets for the four atoms within the hexagonal P unit cell you have drawn in (a). Express the coordinates of atoms in units of the lattice constants a and c.

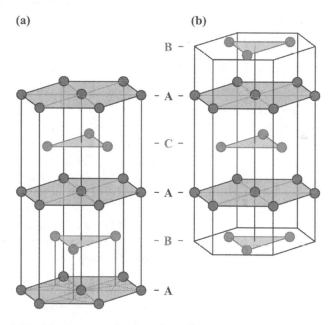

Figure 3.35 Two hexagonal prisms for a *dhcp* structure. One is shifted with respect to the other by 1/4 of the prism height.

Exercise 3.7 Gadolinium at room temperature and 44 GPa crystallizes in the *triple hexagonal close-packed* (*thcp*) structure, which is a six-layered structure with an **ABCBACABCBAC**... layer sequence.

(a) Draw the hexagonal prism for this structure showing the layer sequence.

(b) Within the hexagonal prism, draw the hexagonal P unit cell of the *thcp* structure with the basis vectors $\vec{a}$, $\vec{b}$, and $\vec{c}$ that define it.

(c) Place the coordinate triplets of the atoms within the hexagonal P unit cell you have drawn in (b). Express the coordinates of atoms in units of the lattice constants a and c.

(d) The experimental cell parameters of Gd at room temperature and 44 GPa are $a = 2.910$ Å and $c = 14.31$ Å. Show that the crystal structure of Gd is nearly an ideal close-packed structure.

(e) Show that each atom has its 12 NNs almost at the same distance to it, which confirms the result obtained in (d).

Exercise 3.8 Samarium under normal conditions (α-Sm) crystallizes in a rhombohedral type structure ($hR3$-Sm):

$$\alpha\text{-Sm structure} \equiv rhombohedral\ lattice + 3\text{-atom basis}.$$

Figure 3.36 shows the hexagonal prism for α-Sm.

(a) Draw the rhombohedral P unit cell of the α-Sm structure inside the hexagonal prism shown in Fig. 3.36.

(b) On a separate draw, show a triple hexagonal R unit cell of the α-Sm structure. What is the ratio between the volume of this cell and the volume of the rhombohedral P unit cell?

(c) How close is the α-Sm structure to an ideal close-packed structure? Express your answer in percentage.

Exercise 3.9 For the α-Sm structure, give the coordinate triplets of the nine atoms within a triple hexagonal R unit cell in obverse setting. Express the coordinates in units of cell parameters a_h and c_h.

Exercise 3.10 The crystal structures of α-As, β-Po, α-Hg, and α-Sm (of $hR2$-As, $hR1$-Po, $hR1$-Hg, and $hR3$-Sm type, respectively) have the rotoinversion axis $\bar{3}$ as an axis of the highest order. The set of atoms shown in Fig. 3.36 for the α-Sm structure has a $\bar{3}$ axis with $\bar{3}$-inversion point at half of the hexagonal prism height. In the case of the β-Po structure, the set of atoms which has a $\bar{3}$ axis is shown in Fig. 3.33c (bottom) for selenium at high pressure. A similar set of atoms has a $\bar{3}$ axis in the case of the α-Hg structure. Starting from

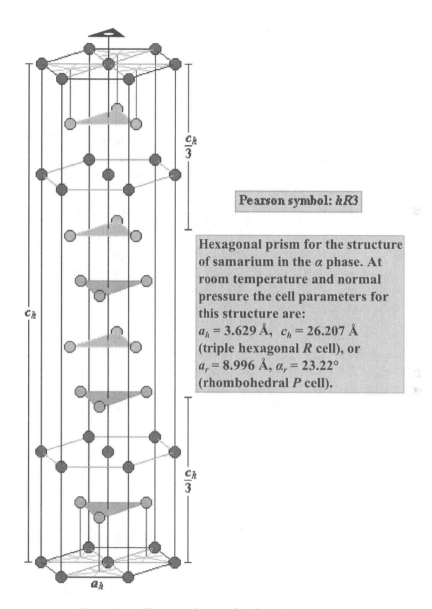

Pearson symbol: *hR3*

Hexagonal prism for the structure of samarium in the α phase. At room temperature and normal pressure the cell parameters for this structure are:
$a_h = 3.629$ Å, $c_h = 26.207$ Å (triple hexagonal R cell), or
$a_r = 8.996$ Å, $\alpha_r = 23.22°$ (rhombohedral P cell).

Figure 3.36 Hexagonal prism for the α-Sm structure.

the set of atoms shown in Fig. 3.32c for α-As, propose a set of atoms which possesses a $\bar{3}$ axis. Show, on this axis, the position of the $\bar{3}$-inversion point.

Exercise 3.11 In the case of ytterbium the transition from α to β phase occurs in a broad temperature range near the room temperature. Tables 3.3 and 3.12 report the experimental lattice constants obtained for both phases at room temperature and normal pressure. Show that, at ambient conditions, the average NN interatomic distance in α-Yb differs from the NN interatomic distance in β-Yb by only about 0.3%.

Exercise 3.12 For cerium the transition from β to γ phase occurs in a broad temperature range near the room temperature. Tables 3.3 and 3.4 report the experimental lattice constants obtained for both phases under normal conditions. Show that, at normal conditions, the average NN interatomic distance in β-Ce differs from the NN interatomic distance in γ-Ce by only about 0.4%.

Exercise 3.13 Compare the NN interatomic distances in α-Fe at normal conditions and in δ-Fe at 1712 K and normal pressure, both having the *bcc* structure. The appropriate lattice constants should be taken from Table 3.2. Note that the volume of a solid usually increases with temperature, and this is reflected by the positive value of the so-called coefficient of thermal expansion.

Exercise 3.14 On the example of metals for which we reported the experimental data for different phases, show that the NN interatomic distance derived from the experimental lattice constant of the *bcc* structure is smaller than the NN interatomic distance obtained from the data reported for close-packed structures, although the data for the *bcc* structure were obtained at higher temperatures. To do so, compare the NN interatomic distances for the following cases:

(a) β-Ca (at 773 K) and α-Ca (at room temperature),
(b) δ-Ce (at 1030 K) and γ-Ce (at room temperature),
(c) δ-Fe (at 1712 K) and γ-Fe (at 1373 K),
(d) β-Li (at room temperature) and α-Li (at 78 K),
(e) β-Na (at room temperature) and α-Na (at 5 K).

The appropriate lattice constants are listed in Tables 3.2, 3.3, and 3.12.

Exercise 3.15 The NNs, NNNs, TNNs, fourth nearest neighbors, and also half of the fifth nearest neighbors of an atom in the *fcc* structure belong entirely or partially to a hexagonal prism of side $2a_h$ and height $2c_h$, where a_h and c_h are parameters of the triple hexagonal unit cell of the *fcc* structure. The atom in consideration is placed at the geometric center of the large hexagonal prism. Figure 3.37 shows the cross section in the middle of this prism and a gray colored hexagon which corresponds to the bottom base of a hexagonal prism of side a_h and height c_h. The draw also shows up to fifth nearest neighbors (or their projections) of the central atom if such neighbor belongs to the top half of the large prism. The fraction nearby the

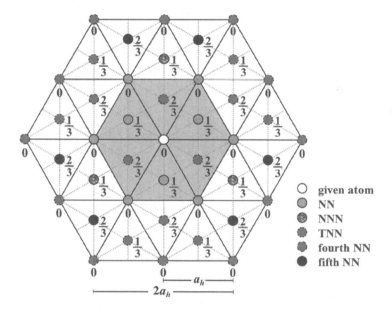

Figure 3.37 Cross section in the middle of a hexagonal prism of side $2a_h$ and height $2c_h$, where a_h and c_h are parameters of a triple hexagonal unit cell of the *fcc* structure. The gray colored hexagon corresponds to the bottom base of the hexagonal prism of side a_h and height c_h. The atoms located in the cross sectional plane and the projections of atoms within the top half of the large prism are shown. The fraction nearby the projection of an atom represents its coordinate in units of c_h.

projection of an atom represents its coordinate in units of c_h. Note that for each neighbor, whose projection is shown on the cross section, there is another neighbor with a negative coordinate.

(a) Calculate the distances from the central atom to its NNs, NNNs, ..., fifth nearest neighbors. Express the results in terms of a_h.

(b) How many NNs, NNNs, TNNs, and fourth nearest neighbors does an atom in the *fcc* structure have?

Chapter 4

Crystal Structures of Important Binary Compounds

4.1 Introduction

In this chapter, we will consider important structures for binary compounds. As we learned already about the examples of elements, the type of crystal structure or at least the preference for the coordination number depends significantly on the type of bonding between the NNs. Until now, we have discussed the structures of elements, mainly with metallic and covalent bonding. In the case of compounds, however, an important role plays the ionic bonding. In most cases, the bonding is partially ionic and partially covalent. It means that the atoms are partially ionized and the atomic radii depend mainly on the degree of their ionization (oxidation state) and also, although less, on the coordination number. The two types of ions in a binary compound have in general different radii and its crystal structure depends strongly on the cation to anion radius ratio.

Basic Elements of Crystallography (2nd Edition)
Nevill Gonzalez Szwacki and Teresa Szwacka
Copyright © 2016 Pan Stanford Publishing Pte. Ltd.
ISBN 978-981-4613-57-6 (Hardcover), 978-981-4613-58-3 (eBook)
www.panstanford.com

4.2 The Ionic Radius Ratio and the Coordination Number

In this section, we will show the relation between the cation to anion radius ratio, r_+/r_-, and the number of NNs of a cation in a binary compound. The cations are in general smaller than the anions, so the r_+/r_- ratio is, in most cases, smaller than 1. The cation tries to surround itself with as many anions as possible and as closely as possible. The packing arrangement in most cases is such that the cations, considered hard spheres, are in contact with the anions, while the anions surround each cation without touching one another. Depending on the r_+/r_- ratio this can be achieved in different arrangements of ions, corresponding to different coordination numbers. Here, we will find the limiting radius ratio for the case of coordination number 4, on the example of the zinc blende structure.

In the zinc blende structure (also known as the sphalerite structure) crystallize binary compounds in which the contribution of covalent bonding to the interatomic bonds is important. Among them there is zinc sulfide in the β phase (β-ZnS), which gives the name to this structure. In β-ZnS each ion has four NNs, which is a characteristic of covalent bonding in binary compounds. Both, Zn and S, have their NNs at the vertices of a regular tetrahedron with Zn or S in its center. This is shown in Fig. 4.1.

It is easy to realize that the zinc blende structure has the same atomic arrangement as the diamond structure, but now the two *fcc*

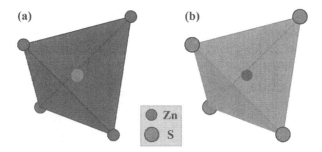

(a)   (b)

Zn
S

Figure 4.1 Zinc blende structure. (a) Regular tetrahedron defined by Zn cations with the S anion in its center. (b) Nearest neighbors of the Zn cation at the vertices of a regular tetrahedron.

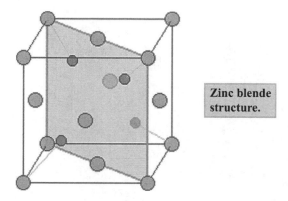

Zinc blende structure.

Figure 4.2 Cubic F unit cell for the zinc blende structure. A plane defined by two body diagonals of the cube is shown.

substructures are made of different ions. The cations occupy half of the tetrahedral interstices in the *fcc* anion substructure and *vice versa*. In Fig. 4.2, we show the cubic F unit cell for the zinc blende structure. In this figure, we have also shown the diagonal cross section of the cube in which we can find the centers of the NNs and the points of contact between them. As an example, the cross section for silicon carbide in the β phase (β-SiC) is shown in Fig. 4.3. Silicon carbide is a IV–IV compound, so it has a large covalent component in its bonds. Therefore, in Fig. 4.3, we have drawn the circles, which represent the cross sections of Si and C atoms, with radii having the same ratio as for the covalent radii of the Si and C elements. The points of contact between neighboring atoms are found on the diagonals of the cube. We can observe in Fig. 4.3 that the Si atoms surround C atoms without touching one another. This is the typical situation in any zinc blende structure. The limiting case is achieved when the anions touch one another. This is shown in Fig. 4.4.

We see in Fig. 4.4 that the sum of the ionic radii is $1/4$ of the body diagonal longitude ($\sqrt{3}a$, where a is the cube edge), therefore

$$r_- + r_+ = \frac{1}{4}\sqrt{3}a. \tag{4.1}$$

On the other hand, r_- is $1/4$ of the length of the cube face diagonal

$$r_- = \frac{1}{4}\sqrt{2}a, \tag{4.2}$$

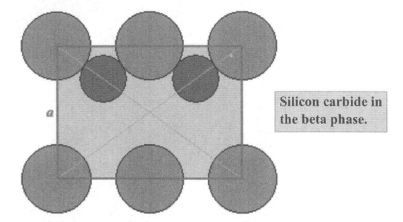

Silicon carbide in the beta phase.

Figure 4.3 Cross section from Fig. 4.2 of the cubic F unit cell for silicon carbide in the β phase. Larger circles correspond to the cross sections of Si atoms and the smaller ones to the cross sections of C atoms.

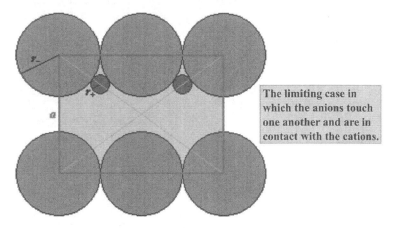

The limiting case in which the anions touch one another and are in contact with the cations.

Figure 4.4 A plane defined by two body diagonals of the cubic F unit cell for the zinc blende structure shown in Fig. 4.2. In the figure, we show the limiting case in which the anions, represented by larger circles, touch one another and are in contact with the cations (smaller circles).

so

$$\begin{cases} r_+ = \dfrac{1}{4}\sqrt{3}a - r_- = \dfrac{1}{4}\left(\sqrt{3} - \sqrt{2}\right)a \\ r_- = \dfrac{1}{4}\sqrt{2}a \end{cases} \tag{4.3}$$

and the radius ratio for the limiting case depicted in Fig. 4.4 is

$$\frac{r_+}{r_-} = \frac{\sqrt{3} - \sqrt{2}}{\sqrt{2}} = \frac{1}{2}\sqrt{6} - 1 \cong 0.225. \qquad (4.4)$$

It is obvious that only in cases when

$$\frac{r_+}{r_-} \geq 0.225 \qquad (4.5)$$

the cations are in contact with anions, otherwise a cation would occupy the central region of the tetrahedral interstice present in the anion substructure, without touching the anions. This situation hardly occurs as the structure would not be stable.

In Table 4.1 we list the limiting radius ratios for different cation coordination numbers. This ratio for the coordination number 4 has been calculated above and the limiting radius ratios for the coordination numbers 6 and 8 will be calculated later. In Table 4.1 we also show the range for the radius ratio that would be expected for each coordination number and the possible crystal structures in which the cations have this coordination number. The ranges for the radius ratios are determined based on the fact that when r_+/r_- reaches the limiting value for the higher coordination number, the structures, in which the cation has this coordination number, become more stable. In practice only about 50% of cases can be classified according to the radius ratio ranges given in Table 4.1. This will be shown on the examples of alkali halides that crystallize in the NaCl structure.

We are assuming in Table 4.1 that the r_+/r_- ratio is less than 1, which means that the cation is smaller than the anion, as is the case in most compounds. In these cases, the cations, which occupy

Table 4.1 Expected radius ratio ranges for different cation coordination numbers. The crystal structures from the last column of the table will be fully described in this chapter

Cation coordination number	Limiting value for r_+/r_-	Expected radius ratio range	Possible crystal structures
4	0.225	0.225–0.414	zinc blende, wurtzite, anti-fluorite
6	0.414	0.414-0.732	sodium chloride, nickel arsenide
8	0.732	0.732-0.999	cesium chloride, fluorite
12	1		

Table 4.2 Ionic radii for Na^+, K^+, and Ca^{2+}, for different coordination numbers. For comparison, we have also listed the metallic radii for Na, K, and Ca taken from Table 3.17

Element	Ionic radius (in Angstroms)				Metallic radius (in Angstroms)
	Coordination number				
	IV	VI	VIII	XII	XII
Na^+	0.99	1.02	1.18	1.39	
Na					1.91
K^+	1.37	1.38	1.51	1.64	
K					2.37
Ca^{2+}		1.00	1.12	1.34	
Ca					1.98

the interstices present in the anion substructure, are expected to touch the anions, which can be achieved in a structure for which the r_+/r_- ratio is larger than the limiting radius ratio for this structure. However, in occasions the situation is the opposite, the cations are larger than the anions, and then the r_-/r_+ ratio has to be considered in the way as r_+/r_- was in Table 4.1. This will be shown later, on the examples of some alkali halides.

We have already mentioned before that the ionic radii depend both on the degree of ionization of the atom and on the coordination number. The dependence on the coordination number is exemplified in Table 4.2. The comparison between the ionic radii and the metallic radius is also done in this table. We can observe the large difference between the values for the metallic and the ionic radii. We can also observe that, in the case of common coordination numbers for binary compounds (IV, VI, and VIII), exists a quite large difference between the values for r_+ in the cases of coordination numbers IV and VI respect to the case of coordination number VIII.

4.3 Zinc Blende Structure

Pearson symbol: *cF*8, **prototype: ZnS.** In Fig. 4.5, we show two types of unit cells for the zinc blende structure. The top part of this figure (Fig. 4.5a) shows two cubic *F* cells, one defined by S anions

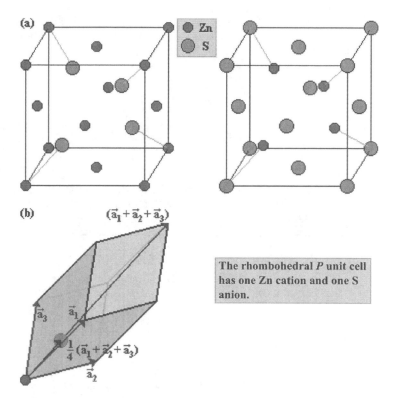

Figure 4.5 (a) Two cubic F unit cells for the zinc blende structure of ZnS: one with Zn cations and the other one with S anions at the vertices. (b) A rhombohedral P unit cell for the zinc blende structure with the two ions belonging to it.

and the other one defined by Zn cations. Both cells are conventional. In addition, in Fig. 4.5b, we show a rhombohedral P unit cell with two ions (one of each kind) belonging to it. So, the zinc blende structure can be considered a *fcc* Bravais lattice with two-atom basis.

In the zinc blende structure crystallize compounds in which the covalent contribution to the bonds prevails over the ionic contribution or at least is significant. Among them there are III–V compounds, for which we have listed the experimental lattice constants in Table 4.3. In this structure crystallize also compounds that contain a transition metal (TM) and an element from columns VI or VII of the periodic table. The lattice constants for those

Table 4.3 Lattice constants (in Angstroms) obtained under normal conditions for III–V compounds that crystallize in the zinc blende structure

	N	P	As	Sb
B	BN (3.6159)	BP (4.5383)	BAs (4.777)	
Al		AlP (5.4625)	AlAs (5.656)	AlSb (6.1355)
Ga	GaN (4.511)	GaP (5.4504)	GaAs (5.65317)	GaSb (6.0961)
In		InP (5.847)	InAs (6.05836)	InSb (6.4794)

Table 4.4 Lattice constants (in Angstroms) obtained under normal conditions for compounds of Be-VI and TM-VI type that crystallize in the zinc blende structure

	O	S	Se	Te	Po
Be		BeS (4.8624)	BeSe (5.1477)	BeTe (5.6225)	BePo (5.838)
Mn		β-MnS (5.601)	β-MnSe (5.902)	α-MnTe (6.338)	
Zn	ZnO (4.63)	β-ZnS (5.4109)	ZnSe (5.6676)	β-ZnTe (6.1037)	ZnPo (6.309)
Cd		β-CdS (5.8304)	CdSe (6.077)	CdTe (6.4809)	CdPo (6.665)
Hg		β-HgS (5.8537)	α-HgSe (6.0854)	α-HgTe (6.453)	

Table 4.5 Lattice constants (in Angstroms) obtained under normal conditions for compounds of TM-VII type that crystallize in the zinc blende structure

	F	Cl	Br	I
Cu	CuF (4.255)	γ-CuCl (5.4202)	γ-CuBr (5.6955)	γ-CuI (6.05844)
Ag				γ-AgI (6.4991)

compounds are given in Tables 4.4 and 4.5 for elements from columns VI and VII, respectively. Four II–VI compounds also crystallize in the zinc blende structure, although most of them, as we will see later, crystallize in the NaCl structure. They are BeS, BeSe, BeTe, and BePo, and the lattice constants for these compounds are listed in Table 4.4.

It was already mentioned in the previous section that silicon carbide in the β phase (β-SiC) also crystallizes in the zinc blende structure. The lattice constant for this compound at normal conditions is 4.35845 Å. In Fig. 4.3, we had a plane defined by two body diagonals of the cubic unit cell for β-SiC, with the cross sections of the Si and C atoms drawn with the radii that have the same ratio as the ratio of the covalent radii for Si and C elements. It is interesting to mention that although silicon carbide has $\sim$12% of ionic contribution to its bonds, the sum of Si and C covalent radii (taken from Table 3.15 in Section 3.17), which is $r_{Si} + r_C = 1.948$ Å, is to within 3% equal to the sum of the Si and C radii obtained using Eq. 4.1. This means that the amount of the ionic character of the bonds is almost not reflected in the sum of the Si and C radii. This result is a consequence of the fact that if the ionic contribution to the bond changes the radius of one atom by Δr the change in the radius of the NN atom is by about $-\Delta r$, so the sum of ionic radii is in a good approximation equal to the sum of covalent radii of these atoms.

In the diamond and zinc blende structures only half of the tetrahedral interstices present in the cubic F unit cell are occupied with atoms or ions. We will show below an example of a structure that has also a cubic F unit cell but with all 8 tetrahedral interstices occupied with ions. This is the case of the calcium fluoride structure.

4.4 Fluorite and Anti-Fluorite Structures

4.4.1 *Fluorite Structure*

Pearson symbol: *cF*12, prototype: CaF$_2$. The calcium fluoride (CaF$_2$) structure, more commonly known as the fluorite structure, has its positive ions forming the *fcc* substructure and usually larger negative ions occupying tetrahedral interstices in this substructure. This is shown in Fig. 4.6a for CaF$_2$. Each F$^-$ anion is placed in the center of a tetrahedral interstice and has 4 NNs (see also Fig. 4.7a). In Fig. 4.6b, we show the cubic F unit cell for the CaF$_2$ structure with the anions in its vertices. We can see in this figure that the anions define 8 small cubes. Four of them have cations in their centers; therefore each Ca^{2+} cation, contrary to the anion, has 8 NNs, as shown in Fig. 4.7b. The unit cell from Fig. 4.6a is conventional.

In addition to CaF$_2$, other II–VII compounds crystallize in the fluorite structure. They are listed at the top of Table 4.6. Among

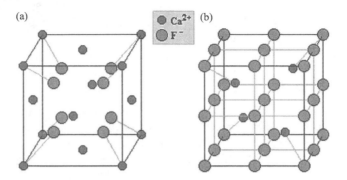

Figure 4.6 Cubic F unit cells for the CaF_2 structure. In the cube vertices are placed Ca^{2+} cations in (a) and F^- anions in (b).

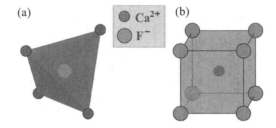

Figure 4.7 (a) Regular tetrahedron defined by the NNs of the F^- anion in CaF_2. (b) Cube defined by the NNs of the Ca^{2+} cation in CaF_2.

other examples of compounds that crystallize in this structure, we can mention hydrides, silicides, oxides, and fluorides of some TMs (mainly RE metals and actinides) and also lead difluoride in the β phase (β-PbF_2) and polonium dioxide in the α phase (α-PoO_2). The experimental lattice constants, obtained under normal conditions for the compounds specified above, are listed in Table 4.6. This table allows to identify quickly which metals form compounds within a given group of hydrides, silicides, oxides, fluorides, or chlorides (see columns of the table) and also allows to see how many compounds with the fluorite structure can be formed by a given metal (see rows of the table).

When the ionic positions are reversed, and the anions and cations occupy the Ca^{2+} and F^- positions, respectively, we obtain the anti-fluorite structure, which will be considered below.

Table 4.6 Lattice constants (in Angstroms) obtained under normal conditions for II–VII compounds and hydrides, silicides, oxides, and fluorides of some TMs, all of them crystallizing in the fluorite structure. In addition, the data for β-PbF$_2$ and α-PoO$_2$ are included

	H	Si	O	F	Cl
Ca				CaF$_2$ (5.46295)	
Sr				SrF$_2$ (5.7996)	SrCl$_2$ (6.9767)
Ba				BaF$_2$ (6.1964)	
Ra				RaF$_2$ (6.368)	
Sc	ScH$_2$ (4.78315)				
Co		CoSi$_2$ (5.365)			
Ni		NiSi$_2$ (5.406)			
Y	YH$_2$ (5.207)				
Zr			ZrO$_2$ (5.09)		
Nb	NbH$_2$ (4.566)				
Cd				CdF$_2$ (5.393)	
Pt	PtH$_2$ (5.517)				
Hg				HgF$_2$ (5.5373)	
Ce	CeH$_2$ (5.581)		CeO$_2$ (5.413)		
Pr	PrH$_2$ (5.516)		PrO$_2$ (5.392)		
Nd	NdH$_2$ (5.4678)				
Sm	SmH$_2$ (5.3773)				
Eu				EuF$_2$ (5.796)	
Gd	GdH$_2$ (5.303)				
Tb	TbH$_2$ (5.246)		TbO$_2$ (5.213)		
Dy	DyH$_2$ (5.2049)				
Ho	HoH$_7$ (5.165)				
Er	ErH$_2$ (5.1279)				
Tm	TmH$_2$ (5.0915)				
Lu	LuH$_2$ (5.0330)				
Th			ThO$_2$ (5.5997)		
Pa			PaO$_2$ (5.505)		
U			UO$_2$ (5.470)		
Np			NpO$_2$ (5.4341)		
Pu			PuO$_2$ (5.39819)		
Am			AmO$_2$ (5.3746)		
Cm			CmO$_2$ (5.368)		
Pb				β-PbF$_2$ (5.9463)	
Po			α-PoO$_2$ (5.637)		

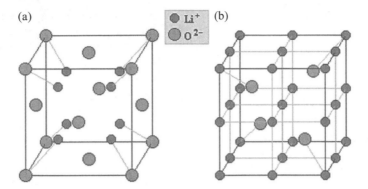

Figure 4.8 Cubic F unit cells of Li_2O which crystallizes in the anti-fluorite structure. In the cube vertices are placed O^{2-} anions in (a) and Li^+ cations in (b).

4.4.2 *Anti-Fluorite Structure*

Pearson symbol: $cF12$, prototype: Li_2O. In the anti-fluorite structure, the anions are in a *fcc* arrangement and the cations occupy all the tetrahedral interstices present in the anion substructure. This is shown in Fig. 4.8a. The cations have a coordination number 4. Figure 4.8b shows the cubic F unit cell for the anti-fluorite structure with the cations at the vertices. We can observe in that figure that the 8 NNs of an anion are placed at the vertices of a small cube that represents one eight of the cubic F unit cell. The NNs of a cation and an anion are shown in Figs. 4.9a and 4.9b, respectively.

In the anti-fluorite structure crystallize some alkali metals with elements from column VI of the periodic table, forming metal oxides,

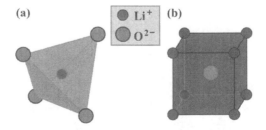

Figure 4.9 (a) Regular tetrahedron defined by the NNs of the Li^+ cation in Li_2O. (b) Cube defined by the NNs of the O^{2-} anion in Li_2O.

Table 4.7 Lattice constants (in Angstroms) obtained under normal conditions for I–VI compounds that crystallize in the anti-fluorite structure

	O	S	Se	Te
Li	Li_2O (4.6114)	Li_2S (5.71580)	Li_2Se (6.0014)	Li_2Te (6.517)
Na	Na_2O (5.55)	Na_2S (6.5373)	Na_2Se (6.825)	Na_2Te (7.314)
K	K_2O (6.436)	K_2S (7.406)	K_2Se (7.676)	K_2Te (8.152)
Rb	Rb_2O (6.755)	Rb_2S (7.65)		

Table 4.8 Lattice constants obtained under normal conditions for some II–III and II–IV compounds, and also phosphides, all of them crystallizing in the anti-fluorite structure

Compound	$a(\text{Å})$	Compound	$a(\text{Å})$
Be_2B	4.663	Mg_2Sn	6.765
Be_2C	4.3420	Mg_2Pb	6.815
Mg_2Si	6.351	Rh_2P	5.5021
Mg_2Ge	6.3894	Ir_2P	5.543

sulfides, selenides, and tellurides. They all are listed in Table 4.7. In Table 4.8 we list some II–III and II–IV compounds, and also phosphides of TMs that crystallize in the anti-fluorite structure.

4.5 Wurtzite Structure

Pearson symbol: *hP*4, prototype: ZnS. Zinc sulfide and most of the binary compounds that crystallize in the zinc blende structure crystallize also in a hexagonal structure, the so-called wurtzite structure. ZnS in the wurtzite structure is in the α phase (α-ZnS). The wurtzite structure is composed of two-dimensional hexagonal layers **A** and **B** and is of the **AABBAABB**...type, where one layer (**A** or **B**) corresponds to one kind of ions and another one to the other kind of ions, so in the case of α-ZnS we have

$$A_{Zn}A_S B_{Zn}B_S A_{Zn}A_S B_{Zn}B_S \ldots,$$

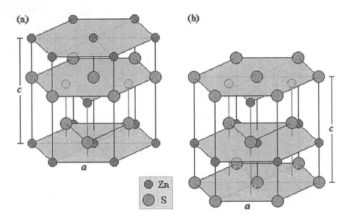

Figure 4.10 Hexagonal prism for ZnS in the wurtzite structure, with Zn cations at the vertices (a) and with S anions at the vertices (b).

whereas in the case of β-ZnS (ZnS in the zinc blende structure) we have

$$A_{Zn}A_SB_{Zn}B_SC_{Zn}C_SA_{Zn}A_SB_{Zn}B_SC_{Zn}C_S....$$

In Fig. 4.10 we show two hexagonal prisms for α-ZnS: one with Zn cations at the vertices (see Fig. 4.10a) and the other one with S anions at the vertices (see Fig. 4.10b). In this figure it is also easy to distinguish the two substructures of the wurtzite structure: that formed by cations and that formed by anions. Each substructure is a *hcp* structure. However, the ions in it do not touch each other, since the NNs of an ion in the wurtzite structure are of another type. Each ion from one substructure occupies a tetrahedral interstice from the other substructure.

The smallest volume that can reproduce the wurtzite structure is the hexagonal unit cell. Figure 4.11 shows two hexagonal P cells for the wurtzite structure of ZnS, one with a Zn cation at the origin (see Fig. 4.11a) and the other one with a S anion at the origin (see Fig. 4.11b). In both cases, the origins are non-conventional (the conventional origin lies on a sixfold screw axis 6_3 shown in Fig. A.30, Appendix). The hexagonal P unit cell for the wurtzite structure contains two ions of each type. We have shown in Fig. 4.12 the coordinate triplets for the four ions belonging to each unit cell from Fig. 4.11. The coordinates are expressed in units of a and c.

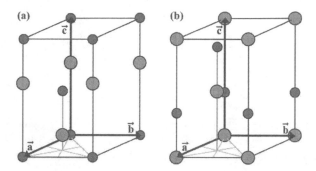

Figure 4.11 Two hexagonal P unit cells for ZnS in the wurtzite structure: in (a) with a Zn cation at the origin and in (b) with a S anion at the origin.

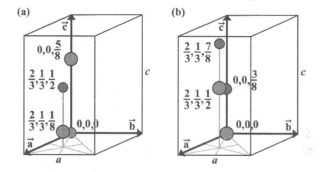

Figure 4.12 (a) and (b) give the coordinate triplets of ions belonging to the unit cells from Figs. 4.11a and 4.11b, respectively. The coordinates are expressed in units of a and c.

Similarly to the zinc blende structure, each ion in the wurtzite structure has a tetrahedral arrangement of the four NNs, although the ionic contribution to their bonds is, in general, larger than the covalent one. This is shown in Figs. 4.13a and 4.13b for α-ZnS, where four NNs surround the S and Zn ions, respectively. These central ions and their NNs can be found inside the hexagonal cells from Fig. 4.11.

We will consider now an ideal case, when the tetrahedrons from Fig. 4.13 are regular. The parameters of the hexagonal unit cell, a and c, fulfill then the relation $c/a = \sqrt{8/3} \cong 1.633$, as in the case of an ideal *hcp* structure. The wurtzite structure of ZnS and many other binary compounds is very close to the ideal case. This

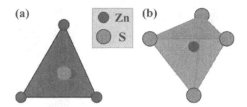

Figure 4.13 (a) A tetrahedron defined by the NNs of the S anion in α-ZnS. (b) A tetrahedron defined by the NNs of the Zn cation in α-ZnS. We can envision the tetrahedrons from (a) and (b) in both cells from Fig. 4.11.

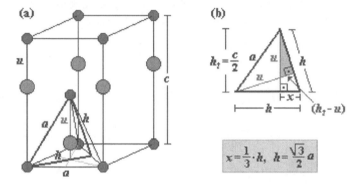

Figure 4.14 (a) A regular tetrahedron, defined by Zn cations, located inside a hexagonal P unit cell for the wurtzite structure of ZnS. (b) A vertical cross section of the tetrahedron shown in (a). See text for detailed explanation.

can be seen in Table 4.9 where we list, in the last column of that table, the c/a ratios for compounds that crystallize in the wurtzite structure. In Fig. 4.14a, we show a regular tetrahedron defined by Zn cations, which is inside a hexagonal unit cell. The cations from the tetrahedron vertices are the NNs of the S anion located in the center of the tetrahedron. We can also see in Fig. 4.14a that the distance, u, between NNs defines also the distance between layers A_S and A_{Zn}, so the S hcp substructure is shifted with respect to the Zn substructure by u along the c axes.

Let us now express u as a function of the lattice parameters. In Fig. 4.14a, we show a vertical cross section of the regular tetrahedron that includes one of its edges a and two heights h of the tetrahedron faces, which are equilateral triangles. The three segments (a, h, h) define a triangle shown in Fig. 4.14b. Inside this triangle we highlighted a right triangle of sides $h_t - u$, $h - x$, and u. There is also a larger triangle that is similar to the highlighted one and have sides x, h, and h_t. The lengths of x is $h/3$. From the similitude of the last two triangles we have

$$\frac{h_t - u}{u} = \frac{x}{h} = \frac{\frac{1}{3}h}{h} = \frac{1}{3}, \tag{4.6}$$

then

$$h_t - u = \frac{1}{3}u, \tag{4.7}$$

and finally

$$u = \frac{3}{4}h_t \underset{h_t = c/2}{\Rightarrow} u = \frac{3}{8}c. \tag{4.8}$$

So, in an ideal wurtzite structure each atom has 4 NNs at a distance $(3/8)\,c$. As it was mentioned before, the same distance can be found between layers A_S and A_{Zn} and, of course, also layers B_S and B_{Zn}. This is shown in Fig. 4.15 for the wurtzite structure of ZnS. In this figure, we show also that the distance between layers A_{Zn} and B_S or B_{Zn} and A_S is $(1/8)\,c$.

Each anion from layers A_S or B_S has 4 NNs located in adjacent A_{Zn} and B_{Zn} layers at a distance $(3/8)\,c$ (see Fig. 4.15), since the c/a

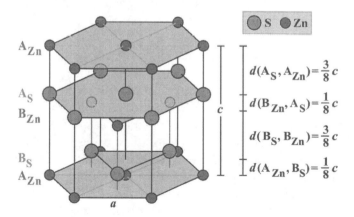

Figure 4.15 Hexagonal layers A_S, A_{Zn}, B_S, and B_{Zn} in the wurtzite structure of ZnS. The distances between the consecutive layers for the ideal case are shown.

Table 4.9 Lattice parameters, obtained under normal conditions, of binary compounds that crystallize in the wurtzite structure

Compound	$a(\text{Å})$	$c(\text{Å})$	c/a
CuH	2.893	4.614	1.59
α-BeO	2.6967	4.3778	1.62
γ-MnS	3.987	6.438	1.61
γ-MnSe	4.12	6.72	1.63
γ-MnTe	4.48	7.32	1.63
γ-ZnO	3.25030	5.2072	1.60
α-ZnS	3.8227	6.2607	1.64
ZnSe	4.003	6.540	1.63
γ-ZnTe	4.31	7.09	1.65
α-CdS	4.1365	6.7160	1.62
CdSe	4.2999	7.0109	1.63
β-AgI	4.599	7.524	1.64
BN	2.555	4.21	1.65
AlN	3.11197	4.98089	1.60
GaN	3.1878	5.1850	1.63
InN	3.53774	5.7037	1.61
SiC	3.079	5.053	1.64

ratio is for ZnS close to the case of an ideal wurtzite structure. The NNNs of an anion are 12 anions at a distance a: six from the layer to which belongs the anion in consideration and the other six from two adjacent layers in the substructure of anions. Since the anions do not touch each other, this substructure is not, of course, close-packed. The same analysis is valid for the substructure of Zn cations.

The experimental lattice parameters obtained under normal conditions for binary compounds that crystallize in the wurtzite structure are given in Table 4.9. We can observe that the c/a ratio, which is given in the last column of the table, is for each case close to that for the ideal case and as a consequence, there is a similarity between the hexagonal and cubic structures of these compounds, although the symmetry of both structures is different.

Let us now summarize important similarities and differences between the zinc blende and the ideal wurtzite structures:

(a) Looking at the NNs, we cannot tell whether it is zinc blende or wurtzite structure.

Table 4.10 Comparison between NN distances for zinc blende and wurtzite structures of some binary compounds. The values were obtained from the lattice constants listed in Tables 4.3, 4.4, and 4.9

Compound	NN distance (Å)	
	Zinc blende	Wurtzite
MnSe	2.555	2.52
MnTe	2.744	2.75
ZnSe	2.454	2.453
CdSe	2.631	2.629
GaN	1.953	1.944

(b) In both cases, the NN and the NNN distances are very close in value. The values for the NNs of some compounds are listed in Table 4.10.

(c) The number (12) of NNNs is the same in both cases.

(d) There is a difference in the location of 3 NNNs. This will be explained in details below.

In both structures, zinc blende and wurtzite, each ion has 6 NNNs in the layer, let us say **A**, to which belongs. The other 6 of 12 NNNs belong to two adjacent layers in the substructure of the ion in consideration. In the case of the wurtzite structure, the adjacent layers are of **B** type, while in the case of the zinc blende structure one of them is of **B** type and the other one is of **C** type. Therefore, the 3 NNNs of an atom from layer **A**, that make the difference between the two structures, are located in the **C** layer in the zinc blende structure and in the **B** layer in the wurtzite structure (in both cases in the substructure of the ion in consideration).

4.6 Nickel Arsenide Related Structures

4.6.1 *NiAs Structure*

Pearson symbol: *hP*4, prototype: NiAs. Similarly to the zinc blende and wurtzite structures, the nickel arsenide (NiAs) structure is related to a close-packed arrangement of ions. It is composed of anion and cation layers placed alternately one on top of the

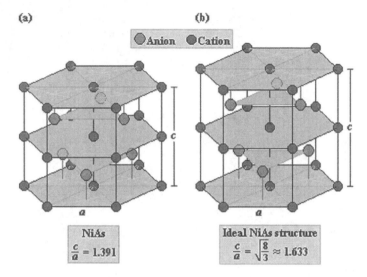

Figure 4.16 (a) Hexagonal prisms for the prototypical NiAs. (b) The NiAs structure for the ideal case when $c/a = \sqrt{8/3}$.

other, in the way illustrated in Fig. 4.16. Each layer represents a two-dimensional hexagonal structure. In Fig. 4.16a, we show the hexagonal prism for the NiAs compound. In addition, we show in Fig. 4.16b the NiAs structure in the ideal case when $c/a = \sqrt{8/3}$. Such case has been already discussed before for the *hcp* and wurtzite structures. We can see in Fig. 4.16 that the c/a ratio for the prototypical NiAs differs 15% from the value that corresponds to the ideal case, while in compounds that crystallize in the wurtzite structure it nearly approaches the ideal ratio.

We can observe in Fig. 4.16 that 6 NNs of a cation located in the center of the hexagonal prism define an octahedron. This octahedron is shown in Fig. 4.17a for the NiAs compound. In the case of the ideal NiAs structure the octahedron is a regular polyhedron and is shown in Fig. 4.17b. In Fig. 4.17 we can also see two additional ions that are the closest cations to the cation placed in the center of each octahedron. In the case of the ideal NiAs structure, those cations are at a distance 15% longer than the distance to the NNs from the cation in consideration. However, in the case of the NiAs compound the distance from a cation to its nearest cations is only

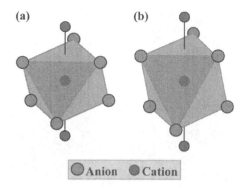

Figure 4.17 Octahedrons defined by anions that are the NNs of a cation (a) in the NiAs compound and (b) in the ideal NiAs structure. In this figure we also show the two nearest cations to the cation placed in the center of each octahedron.

3% longer than the distance to the NNs. It means that each cation in the NiAs compound has effectively 8 NNs (6 anions and 2 cations), all of them forming bonds with this cation (see Fig. 4.17a). It is also important to mention that the length of the Ni-Ni bond in the NiAs compound is, to within 1%, equal to the metallic bond length in the crystal of nickel. Therefore, we can expect that the Ni–Ni bonds, which we are describing here for the NiAs compound, are closer to the metallic bonds than to the ionic ones, since the Ni-Ni ionic bond would be longer than the Ni–Ni metallic bond.

In Fig. 4.18a, we show two hexagonal prisms for the NiAs structure: one with anions and another one with cations at the vertices of the prism. We observe in this figure that each cation from a cation layer lies directly over a cation from any layer below. Therefore, the cations form a simple hexagonal substructure since each of their layers is of the same type. All cation layers are of **C** type (see Fig. 4.18a to the left) and are labeled C_c. In the case of anions, there are two types of layers (labeled A_a and B_a, see Fig. 4.18a to the left) like in the case of the *hcp* structure, therefore the anions form a *hcp* substructure. The two prisms shown in Fig. 4.18a are shifted vertically and horizontally one with respect to the other in the way shown in Fig. 4.18.

Let us now make a comparison between the neighborhood of a cation and an anion in the NiAs structure. We already know

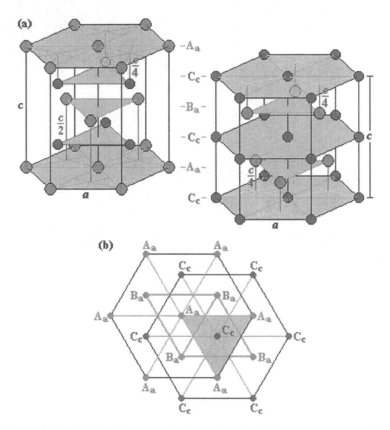

Figure 4.18 (a) Two hexagonal prisms for the NiAs structure: one with anions at the vertices and another one with cations at the vertices. (b) Projection of the centers of ions belonging to each hexagonal prism on the hexagonal base. We can observe that the triangles defined by the ions from the B_a layer have in each case from (a) different orientations with respect to the hexagonal prism base.

that there are tetrahedral and octahedral interstices between consecutive hexagonal layers of different types (see Fig. 3.18, Section 3.12). This is the case of the *hcp* structure. In the case of the wurtzite structure half of the tetrahedral interstices, present in one *hcp* substructure, are occupied by ions belonging to the other substructure and the octahedral interstices are vacant. Also in the case of the zinc blende structure half of the tetrahedral interstices present in the *fcc* substructure of ions of one type are occupied

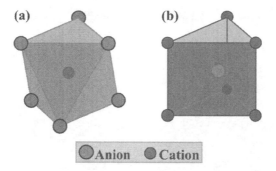

Figure 4.19 Octahedron defined by the NNs of a cation (a) and the trigonal prism defined by the NNs of an anion (b) in the NiAs structure.

by ions of the other type and the octahedral interstices remain vacant. Contrary to those cases, in the NiAs structure the cations occupy all octahedral interstices present in the *hcp* substructure of anions and the tetrahedral interstices are vacant. Turning now to the neighborhood of an anion in the NiAs structure, we can say that all cation layers are of the same type and between them, there are neither octahedral nor tetrahedral interstices. Each anion occupies the center of a trigonal prism, as can be seen in Fig. 4.16.

In Fig. 4.19 we have drawn the neighborhood of the two types of ions in the NiAs structure. Figure 4.19a shows an octahedron with anions in its vertices. These anions are the NNs of a cation that is in the center of the octahedron. Similarly, Fig. 4.19b shows the NNs of an anion that is located in the center of a trigonal prism with cations at the vertices. In both cases the number of NNs is the same but the distribution of cations with respect to the anion is different from the distribution of anions with respect to the cation.

We will now calculate the value for the ideal c/a ratio. This was already done in Section 3.11 for the *hcp* structure. In that opportunity, the calculations were based on the geometric characteristics of a regular tetrahedron defined by the NNs of an atom in the ideal *hcp* structure; the presence of such a tetrahedron inside the hexagonal unit cell of the *hcp* structure determines the c/a ratio in the ideal case. This time, in turn, we will calculate c/a using a regular octahedron defined by the NNs of a cation in the ideal NiAs structure.

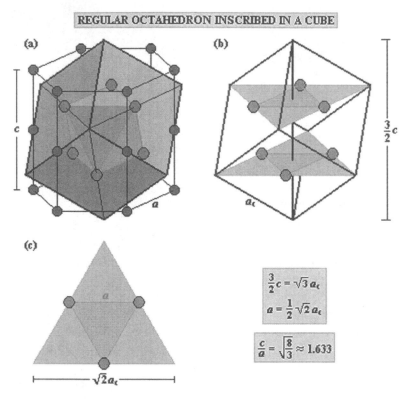

REGULAR OCTAHEDRON INSCRIBED IN A CUBE

(a)

(b)

(c)

$$\frac{3}{2}c = \sqrt{3}\,a_c$$

$$a = \frac{1}{2}\sqrt{2}\,a_c$$

$$\frac{c}{a} = \sqrt{\frac{8}{3}} \approx 1.633$$

Figure 4.20 (a) Hexagonal prism for an ideal NiAs structure with the cations at the vertices. The regular octahedron defined by 6 anions located inside this prism is also shown. In addition, this octahedron is inscribed in a cube. (b) The cube defined in (a). The longitude of a body diagonal of the cube is expressed as a function of the lattice constant c. (c) One of the triangles shown in (b). In this figure we show the relation between the lattice constant a and the cube edge a_c.

As we already know, a regular octahedron may be inscribed in a cube. Figure 4.20 shows such a situation. The longitude of the cube body diagonal is equal to $(3/2)\,c$, as can be verified in the following way:

(a) The anion layers cross the body diagonal of the cube in the points that divide the diagonal in three segments of the same longitude.

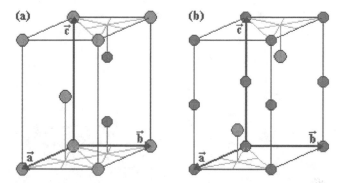

Figure 4.21 Two hexagonal P unit cells for the NiAs structure: (a) with an anion at the origin and (b) with a cation at the origin. Neither of the two choices for the origin is considered standard. In (a) the origin coincides with an $\bar{6}$-inversion point, whereas in (b) it coincides with a center of symmetry placed on the axis $6_3/m$ (see Section 3.11 for the *hcp* structure and Fig. A.31 in Appendix).

(b) The distance between two consecutive anion layers is equal to $c/2$, so, taking into account point (a), we can conclude that the longitude of the body diagonal of the cube is equal to $3\,(c/2) = (3/2)\,c$.

From the considerations made in Fig. 4.20, we can conclude that, indeed, the octahedron defined by the NNs of a cation is a regular polyhedron when $c/a = \sqrt{8/3}$.

The smallest unit cell that can reproduce the NiAs structure is the hP cell defined by anions in Fig. 4.21a or the hP cell defined by cations in Fig. 4.21b. In both cases, the unit cell contains two anions and two cations. Figure 4.22 shows the coordinate triplets of the four ions belonging to each unit cell from Fig. 4.21. The coordinates are expressed in units of a and c.

Some binary compounds crystallize in the so-called anti-NiAs structure that is the same as the NiAs structure, but with cations replaced by anions, and *vice versa*. Figure 4.23 shows the anti-NiAs structure on the example of the VP compound. We can see in this figure that now the vanadium cations form the *hcp* substructure, while the phosphorus anions are arranged in a simple hexagonal substructure.

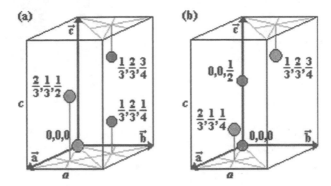

Figure 4.22 (a) and (b) show the coordinate triplets of ions belonging to the unit cells from Figs. 4.21a and 4.21b, respectively. The coordinates are expressed in units of *a* and *c*.

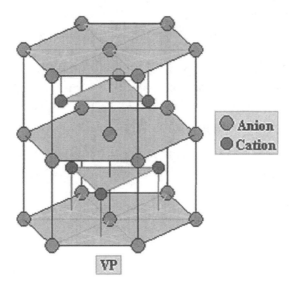

Figure 4.23 Anti-NiAs structure shown on the example of the VP compound.

In the NiAs structure crystallize compounds that contain TMs and elements from columns III, IV, V, or VI of the periodic table. The following compounds may be included:

Column III
B: PtB*
Tl: NiTl

Column IV
C: γ'-MoC**
Sn: FeSn, NiSn, CuSn, RhSn, PdSn, IrSn, PtSn, AuSn
Pb: NiPb, IrPb, PtPb*

Column V
N: δ'-NbN*, ε-NbN**
P: TiP**, VP*, β-ZrP**, HfP**
As: α-TiAs**, β-TiAs*, MnAs, NiAs, α-ZrAs**, HfAs**
Sb: TiSb, VSb, CrSb, MnSb, FeSb, CoSb, NiSb, CuSb, PdSb, IrSb, PtSb
Bi: MnBi*, NiBi*, RhBi*, PtBi*

Column VI:
S: TiS, VS, CrS, β-FeS, CoS, α-NiS, NbS
Se: TiSe, VSe, CrSe, FeSe, CoSe, β-NiSe, RhSe, AuSe
Te: ScTe, TiTe, VTe, CrTe, α-MnTe, FeTe, CoTe, NiTe, ZrTe, RhTe, PdTe, IrTe
Po: MgPo, ScPo*, TiPo, NiPo, ZrPo, HfPo.

In the above classification, the compounds marked with one star crystallize in the anti-NiAs structure, whereas those marked with two stars in the TiAs structure, which will be discussed in the next section.

Experimental lattice parameters for compounds that crystallize in the NiAs, anti-NiAs, or TiAs structures of TM-III, TM-IV, TM-V, and TM-VI type are listed in Tables 4.11, 4.12, 4.13, and 4.14, respectively. We can observe in Tables 4.11 and 4.12 that, with exception of γ'-MoC, all compounds crystallize with the c/a ratio much smaller than the ideal one ($\sqrt{8/3} = 1.633$). In the Table 4.13 more than half of the compounds contain metals from the group of iron. We can also observe in that table that the antimonides and bismuthides of TMs crystallize with the c/a ratio much smaller than the perfect one. In all such cases, the ions that occupy the octahedral interstices in the *hcp* substructure have indeed 8 NNs, like in the case of the Ni cation in NiAs.

Table 4.11 Lattice parameters, obtained under normal conditions, of PtB and NiTl that crystallize in the anti-NiAs and NiAs structures, respectively

Compound	a(Å)	c(Å)	c/a
PtB*	3.358	4.058	1.21
NiTl	4.426	5.535	1.25

*anti-NiAs structure

Table 4.12 Lattice parameters obtained under normal conditions for compounds of TM-IV type that crystallize in the NiAs, anti-NiAs, or TiAs structures

Compound	a(Å)	c(Å)	c/a	Compound	a(Å)	c(Å)	c/a
γ'-MoC**	2.932	10.97	2×1.87	IrSn	3.988	5.567	1.40
γ-FeSn	4.216	5.244	1.24	PtSn	4.104	5.436	1.32
NiSn	4.048	5.123	1.27	AuSn	4.3218	5.523	1.28
CuSn	4.198	5.096	1.21	NiPb	4.15	5.28	1.27
RhSn	4.340	5.553	1.28	IrPb	3.993	5.566	1.39
PdSn	4.378	5.627	1.29	PtPb*	4.258	5.467	1.28

*anti-NiAs structure
**TiAs structure

We can observe in Table 4.14 that the iron group metals are present in more than 2/3 of compounds listed there. The c/a ratios are in this table, for about half of the compounds, quite close to the ideal value, and the CrS, VSe, CrSe, FeSe, ScTe, α-MnTe, and MgPo compounds have the c/a ratio remarkably approaching that value.

Summarizing the data given in Tables 4.11–4.14, we can say that in the case of nickel arsenide related structures the values for c/a are in the wide range between 1.21 and 1.96. As a consequence the ions in these structures may have a different number of NNs and NNNs. Let us see this on the example of a cation in the NiAs structure. For the lower-bound value of c/a each cation has 8 NNs (6 anions and 2 cations) and 6 cations as NNNs, while for the upper-bound value of the c/a ratio a cation in the NiAs structure has 6 anions as NNs and 8 NNNs (all of them cations).

In the next section, we will describe the TiAs structure which is related to the NiAs structure.

Table 4.13 Lattice parameters obtained under normal conditions for compounds of TM-V type that crystallize in the NiAs, anti-NiAs, or TiAs structures

Compound	$a(Å)$	$c(Å)$	c/a	Compound	$a(Å)$	$c(Å)$	c/a
δ'-NbN*	2.968	5.549	1.87	CrSb	4.115	5.493	1.33
ε-NbN**	2.9513	11.248	2×1.91	MnSb	4.140	5.789	1.40
TiP**	3.513	11.75	2×1.67	FeSb	4.072	5.140	1.26
VP*	3.178	6.222	1.96	CoSb	3.866	5.188	1.34
β-ZrP**	3.684	12.554	2×1.70	NiSb	3.9325	5.1351	1.31
HfP**	3.65	12.38	2×1.70	CuSb	3.874	5.193	1.34
α-TiAs**	3.642	12.064	2×1.66	PdSb	4.078	5.593	1.37
β-TiAs*	3.645	6.109	1.68	IrSb	3.978	5.521	1.39
MnAs	3.722	5.702	1.53	PtSb	4.126	5.481	1.33
NiAs	3.619	5.034	1.39	MnBi*	4.290	6.126	1.43
α-ZrAs**	3.804	12.867	2×1.69	NiBi*	4.07	5.33	1.31
HfAs**	3.765	12.680	2×1.68	RhBi*	4.0894	5.6642	1.39
TiSb	4.1033	6.2836	1.53	PtBi*	4.315	5.490	1.27
VSb	4.27	5.447	1.28				

*anti-NiAs structure
**TiAs structure

Table 4.14 Lattice parameters obtained under normal conditions for compounds of TM-VI type that crystallize in the NiAs or anti-NiAs structures. The values for MgPo are also included in the table

Compound	$a(Å)$	$c(Å)$	c/a	Compound	$a(Å)$	$c(Å)$	c/a
TiS	3.299	6.380	1.93	VTe	3.942	6.126	1.55
VS	3.33	5.82	1.75	CrTe	3.978	6.228	1.57
CrS	3.419	5.55	1.62	α-MnTe	4.147	6.711	1.62
β-FeS	3.4436	5.8759	1.71	FeTe	3.800	5.651	1.49
CoS	3.374	5.187	1.54	CoTe	3.888	5.378	1.38
α-NiS	3.4395	5.3514	1.56	NiTe	3.965	5.358	1.35
NbS	3.32	6.46	1.95	ZrTe	3.953	6.647	1.68
TiSe	3.572	6.205	1.74	RhTe	3.987	5.661	1.42
VSe	3.66	5.95	1.63	PdTe	4.152	5.672	1.37
CrSe	3.71	6.03	1.63	IrTe	3.939	5.386	1.37
FeSe	3.62	5.92	1.64	MgPo	4.345	7.077	1.63
CoSe	3.62	5.286	1.46	ScPo*	4.206	6.92	1.65
β-NiSe	3.6613	5.3562	1.46	TiPo	3.992	6.569	1.65
RhSe	3.642	5.486	1.51	NiPo	3.95	5.68	1.44
AuSe	4.12	5.39	1.31	ZrPo	4.031	6.907	1.71
ScTe	4.120	6.748	1.64	HfPo	4.058	6.717	1.66
TiTe	3.834	6.390	1.67				

*anti-NiAs structure

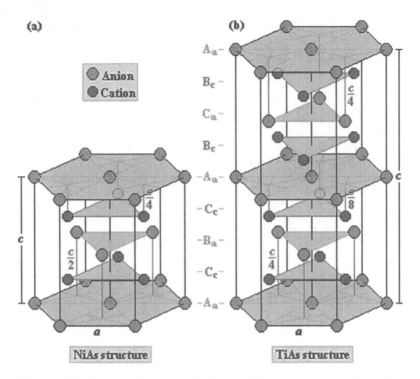

Figure 4.24 Hexagonal prisms: (a) for the NiAs structure and (b) for the TiAs structure. The lattice parameters, *a* and *c*, are shown in both cases.

4.6.2 *TiAs Structure*

Pearson symbol: *hP*8, Prototype: TiAs. In the TiAs structure, the anions are arranged in the *dhcp* substructure shown in Fig. 3.9 (Section 3.8) and the cations occupy all octahedral interstices present in it. We can observe in Fig. 4.24 that the arrangement of ions in the down half of the hexagonal prism for the TiAs structure (see Fig. 4.24b) looks the same as the arrangement of ions in the hexagonal prism for the NiAs structure (see Fig. 4.24a). The sequence of the two-dimensional *hcp* layers in the TiAs structure is the following:

$$A_a C_c B_a C_c A_a B_c C_a B_c A_a C_c B_a C_c A_a B_c C_a B_c \ldots ,$$

where it is easy to separate the layer sequence $A_a B_a A_a$ $C_a A_a B_a A_a C_a \ldots$ corresponding to the *dhcp* anion substructure from the layer sequence $C_cC_cB_cB_cC_cC_cB_cB_c$... for the cation substructure. In the last sequence, we can observe the presence of consecutive cation layers of both the same and different type, which marks the difference from the NiAs structure whose cation substructure has all the layers of the same type. As a consequence, between consecutive layers of the TiAs cation substructure are present not only trigonal prism interstices, but also octahedral and tetrahedral interstices. Half of the trigonal prism interstices and all octahedral interstices are occupied by anions.

4.7 Sodium Chloride Structure

Pearson symbol: $cF8$, prototype: NaCl. We will now talk about the structure of sodium chloride. In those compounds that crystallize in this structure the ionic bonding prevails over the covalent one. Most of the binary compounds that have a high degree of ionicity in their bonds crystallize in this structure and among them the alkali halides which have over 90% of ionic contribution in their bonds.

In alkali halides the positive ion is one of the alkali metals (Li^+, Na^+, K^+, Rb^+, or Cs^+) and the negative ion is one of the halogens (F^-, Cl^-, Br^-, or I^-). Except for CsCl, CsBr, and CsI, all of them crystallize in the NaCl structure under normal conditions.

In the NaCl structure, each ion has six NNs and both, the anion and the cation, have their NNs at the vertices of a regular octahedron with the anion or cation in its center, as shown, using as example the NaCl compound, in Fig. 4.25. The coordination number 6, which is higher than for the case of the zinc blende and wurtzite structures, allows to maximize the ionic bonding.

The sodium chloride structure represents a sequence of two-dimensional hexagonal layers of

$$A_{Na}C_{Cl}B_{Na}A_{Cl}C_{Na}B_{Cl}A_{Na}C_{Cl}B_{Na}A_{Cl}C_{Na}B_{Cl}...$$

type. This sequence can be seen as a superposition of two subsequences of **ABCABC**... type, one for cations and another one for anions. The cation layers are displaced with respect to the anion layers in the way that each ion has 6 NNs.

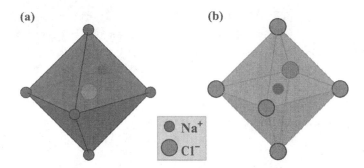

(a)

(b)

● Na^+

◔ Cl^-

Figure 4.25   The structure of NaCl. (a) Regular octahedron defined by Na^+ cations with the Cl^- anion in its center. (b) Nearest neighbors of a Na^+ cation at the vertices of a regular octahedron.

Concluding, we can say that the NaCl structure is a superposition of two *fcc* substructures, each one for a given type of ions. Two conventional cubic *F* unit cells can be proposed for this structure: one defined by anions and the other one defined by cations. These two cells are shown in Fig. 4.26a. In Fig. 4.26b, we show a rhombohedral *P* unit cell with two ions (an anion and a cation) belonging to it, which is the smallest unit cell that reproduces the NaCl structure. Like in the case of the zinc blende structure, the sodium chloride structure can be considered a *cF* Bravais lattice with two-atom basis consisting of one cation and one anion (see Fig. 4.5).

We can observe in Fig. 4.26a that the substructure of anions (cations) is displaced with respect to the substructure of cations (anions) along the cube edge by half of its lengths. Thus, an anion (cation) occupies an octahedral interstice in the cation (anion) substructure. Figure 4.27 illustrates the stacking of **A, B, C** layers for both types of ions in the cubic unit cell of NaCl. In this figure, the $A_{Na}C_{Cl}B_{Na}A_{Cl}C_{Na}B_{Cl}A_{Na}$... sequence of layers is shown. The layers are orthogonal to a body diagonal of the cube. In Fig. 4.28, we show the coordinate triplets of the eight ions belonging to the cubic *F* unit cell shown on the left side of Fig. 4.26a. The coordinates are expressed in units of *a*.

In the next few tables we will list about 300 binary compounds that crystallize in the NaCl structure. This represents a significant

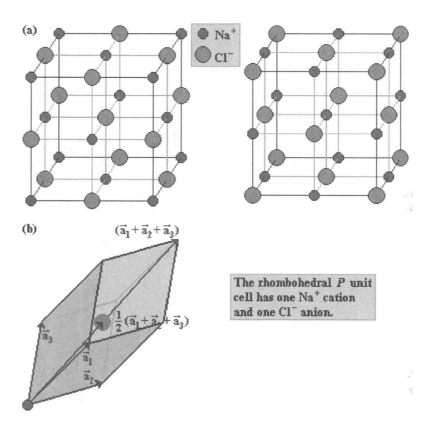

Figure 4.26 (a) Two conventional cubic F unit cells for the structure of sodium chloride: one with Na^+ cations at the vertices and the other one with Cl^- anions at the vertices. (b) A rhombohedral P unit cell with two ions (one anion and one cation), which is the smallest unit cell that reproduces the NaCl structure.

percentage of the total number of compounds having that structure. We begin in Table 4.15 by report experimental lattice constants for I–VII compounds and the silver halides. We can observe in this table that, with exception of CsCl, CsBr, CsI, and AgI, all other I–VII compounds have the NaCl structure. In Table 4.16 are given the lattice constants of II–VI, IV–VI, and V–VI compounds that crystallize in the NaCl structure. We can see in this table that nearly all compounds that contain one of the alkaline earth metals (magnesium, calcium, strontium, and barium) crystallize in this

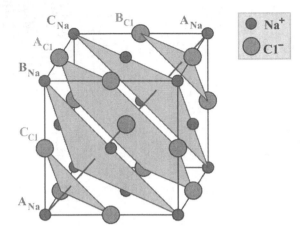

Figure 4.27 The sequence of two-dimensional hexagonal layers A_{Na}, B_{Na}, C_{Na}, A_{Cl}, B_{Cl}, and C_{Cl} in the structure of NaCl.

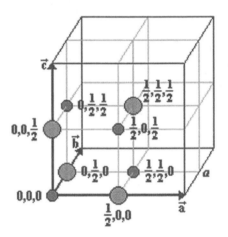

Figure 4.28 Coordinate triplets of ions belonging to the cubic F unit cell of the NaCl structure shown in Fig. 4.26a to the left. The coordinates are expressed in units of a. The cell is conventional.

structure. In Tables 4.17 and 4.18 we report the lattice constants for compounds of TM-VI, and TM-V type, respectively. We can observe in Table 4.17 that nearly all chalcogenides of the RE metals and of the light actinides (thorium, uranium, neptunium, plutonium, and americium) crystallize in the NaCl structure. In this structure also

Table 4.15 Lattice constants (in Angstroms), obtained under normal conditions, of alkali metal halides and silver halides that crystallize in the NaCl structure. In the table, it is also indicated which of the considered compounds crystallize in the CsCl or zinc blende structures

	F	Cl	Br	I
Li	LiF (4.027)	LiCl (5.12952)	LiBr (5.5013)	α-LiI (6.0257)
Na	NaF (4.632)	NaCl (5.6401)	NaBr (5.9732)	NaI (6.4728)
K	KF (5.34758)	KCl (6.2952)	KBr (6.6005)	KI (7.0656)
Rb	RbF (5.6516)	RbCl (6.5810)	RbBr (6.889)	RbI (7.342)
Cs	CsF (6.014)	*CsCl*	*CsCl*	*CsCl*
Ag	AgF (4.92)	AgCl (5.5463)	AgBr (5.7721)	*zinc blende*

Table 4.16 Lattice constants (in Angstroms), obtained under normal conditions, of II-VI and also some IV-VI and V-VI compounds that crystallize in the NaCl structure. In the table, it is also indicated which of the considered compounds crystallize in the wurtzite or NiAs structures

	O	S	Se	Te	Po
Mg	MgO (4.2113)	MgS (5.20182)	MgSe (5.451)	*wurtzite*	*NiAs*
Ca	CaO (4.8105)	CaS (5.6948)	CaSe (5.916)	CaTe (6.356)	CaPo (6.514)
Sr	SrO (5.1615)	SrS (6.0198)	SrSe (6.2432)	SrTe (6.660)	SrPo (6.796)
Ba	BaO (5.539)	BaS (6.3875)	BaSe (6.593)	BaTe (7.0012)	BaPo (7.119)
Sn		SnS (5.80)	SnSe (5.99)	SnTe (6.320)	
Pb		PbS (5.9362)	PbSe (6.1243)	PbTe (6.4591)	PbPo (6.590)
Bi			BiSe (5.99)	BiTe (6.47)	

crystallize oxides of the TMs, which are mainly from the iron group, and oxides of the actinides mentioned above. In Table 4.17 we can also observe that the chalcogenides of TMs that are not RE metals often crystallize in the NiAs structure. Among compounds of the TM-VI type, there is also a small group of compounds that crystallize in the zinc blende structure, like CdPo and cadmium and mercury chalcogenides.

Similarly as in Table 4.17 are organized the experimental data in Table 4.18 for TM nitrides, phosphides, arsenides, antimonides, and bismuthides. As was the case in Table 4.17, in this table the

Table 4.17 Lattice constants (in Angstroms) obtained under normal conditions for compounds of the TM-VI type that crystallize in the NaCl structure. In the table, it is also indicated which of the considered compounds crystallize in the NiAs or zinc blende structures

	O	S	Se	Te	Po
Sc		ScS (5.19)	ScSe (5.398)	*NiAs*	*NiAs*
Ti	TiO (4.1766)	*NiAs*	*NiAs*	*NiAs*	*NiAs*
V	VO (4.073)	*NiAs*	*NiAs*	*NiAs*	
Cr	CrO (4.16)		*NiAs*	*NiAs*	
Mn	MnO (4.446)	α-MnS (5.2236)	α-MnSe (5.462)	*NiAs*	
Fe	FeO (4.326)	*NiAs*	*NiAs*	*NiAs*	
Co	CoO (4.264)	*NiAs*	*NiAs*	*NiAs*	
Ni	NiO (4.1771)	*NiAs*	*NiAs*	*NiAs*	*NiAs*
Y		YS (5.493)	YSe (5.711)	YTe (6.098)	
Zr	ZrO (4.62)	ZrS (5.1522)		*NiAs*	*NiAs*
Nb	NbO (4.212)	*NiAs*			
Rh			*NiAs*	*NiAs*	
Pd				*NiAs*	
Cd	CdO (4.6953)	*zinc blende*	*zinc blende*	*zinc blende*	*zinc blende*
Hf					*NiAs*
Ta	TaO (4.431)				
Ir				*NiAs*	
Pt	PtO (5.15)				
Au			*NiAs*		
Hg		*zinc blende*	*zinc blende*	*zinc blende*	HgPo (6.250)
La		LaS (5.854)	LaSe (6.066)	LaTe (6.429)	
Ce		CeS (5.779)	CeSe (5.9920)	CeTe (6.36)	
Pr		PrS (5.731)	PrSe (5.944)	PrTe (6.315)	
Nd		NdS (5.689)	NdSe (5.907)	NdTe (6.282)	
Sm	SmO (4.9883)	SmS (5.9718)	SmSe (6.202)	SmTe (6.594)	SmPo (6.724)
Eu	EuO (5.142)	EuS (5.9708)	EuSe (6.197)	EuTe (6.594)	EuPo (6.720)
Gd		GdS (5.565)	GdSe (5.76)	GdTe (6.139)	
Tb		TbS (5.5221)	TbSe (5.7438)	TbTe (6.1150)	TbPo (6.254)
Dy		DyS (5.489)	DySe (5.690)	DyTe (6.079)	DyPo (6.214)
Ho		HoS (5.465)	HoSe (5.680)	HoTe (6.049)	HoPo (6.200)
Er		ErS (5.422)	ErSe (5.656)	ErTe (6.063)	
Tm		TmS (5.412)	TmSe (5.688)	TmTe (6.346)	TmPo (6.256)
Yb	YbO (4.86)	YbS (5.687)	YbSe (5.9321)	YbTe (6.361)	YbPo (6.542)
Lu		LuS (5.355)	LuSe (5.572)	LuTe (5.953)	LuPo (6.159)
Th		ThS (5.6851)	ThSe (5.880)		
Pa	PaO (4.961)				
U	UO (4.92)	US (5.486)	USe (5.751)	UTe (6.155)	
Np	NpO (5.01)	NpS (5.527)	NpSe (5.8054)	NpTe (6.2039)	
Pu	PuO (4.958)	PuS (5.5412)	PuSe (5.7934)	PuTe (6.1774)	
Am	AmO (5.045)	AmS (5.592)		AmTe (6.176)	

Table 4.18 Lattice constants (in Angstroms) obtained under normal conditions for the compounds of the TM-V type that crystallize in the NaCl structure. The data for some tin pnictides are also included. In addition, we indicate in the table which of the considered compounds crystallize in the NiAs or TiAs structures

	N	P	As	Sb	Bi
Sc	ScN (4.44)	ScP (5.312)	ScAs (5.487)	ScSb (5.8517)	ScBi (5.954)
Ti	TiN (4.235)	*TiAs*	*NiAs and TiAs*	*NiAs*	
V	VN (4.1361)	*NiAs*		*NiAs*	
Cr	CrN (4.148)			*NiAs*	
Mn			*NiAs*	*NiAs*	*NiAs*
Fe				*NiAs*	
Co				*NiAs*	
Ni			*NiAs*	*NiAs*	*NiAs*
Cu				*NiAs*	
Y	YN (4.877)	YP (5.661)	YAs (5.786)	YSb (6.165)	YBi (6.256)
Zr	ZrN (4.585)	α-ZrP (5.263) and *TiAs*	β-ZrAs (5.4335) and *TiAs*		
Nb	δ-NbN (4.394), *NiAs*, and *TiAs*				
Rh					*NiAs*
Pd				*NiAs*	
Hf	HfN (4.52)	*TiAs*	*TiAs*		
Ta	*NiAs*				
Ir				*NiAs*	
Pt				*NiAs*	*NiAs*
La	LaN (5.301)	LaP (6.0346)	LaAs (6.151)	LaSb (6.490)	LaBi (6.578)
Ce	CeN (5.020)	CeP (5.909)	CeAs (6.072)	CeSb (6.420)	CeBi (6.5055)
Pr	PrN (5.155)	PrP (5.903)	PrAs (6.009)	PrSb (6.375)	PrBi (6.4631)
Nd	NdN (5.132)	NdP (5.838)	NdAs (5.9946)	NdSb (6.321)	NdBi (6.4222)
Sm	SmN (5.0481)	SmP (5.760)	SmAs (5.921)	SmSb (6.271)	SmBi (6.3582)
Eu	EuN (5.017)	EuP (5.7562)			
Gd	GdN (4.9987)	GdP (5.723)	GdAs (5.854)	GdSb (6.217)	GdBi (6.3108)
Tb	TbN (4.9344)	TbP (5.688)	TbAs (5.824)	TbSb (6.178)	TbBi (6.2759)
Dy	DyN (4.9044)	DyP (5.653)	DyAs (5.794)	DySb (6.154)	DyBi (6.2491)
Ho	HoN (4.8753)	HoP (5.626)	HoAs (5.769)	HoSb (6.131)	HoBi (6.228)
Er	ErN (4.842)	ErP (5.606)	ErAs (5.7427)	ErSb (6.106)	ErBi (6.2023)
Tm	TmN (4.8021)	TmP (5.573)	TmAs (5.711)	TmSb (6.087)	TmBi (6.1878)
Yb	YbN (4.7852)	YbP (5.555)	YbAs (5.698)	YbSb (6.079)	
Lu	LuN (4.7599)	LuP (5.533)	LuAs (5.680)	LuSb (6.0555)	LuBi (6.156)
Th	ThN (5.1666)	ThP (5.8324)	ThAs (5.978)	ThSb (6.318)	
Pa			PaAs (5.7560)		
U	UN (4.890)	UP (5.5883)	UAs (5.7767)	USb (6.203)	UBi (6.3627)
Np	NpN (4.897)	NpP (5.6148)	NpAs (5.8366)	NpSb (6.2517)	NpBi (6.370)
Pu	PuN (4.9049)	PuP (5.6613)	PuAs (5.8565)	PuSb (6.2375)	PuBi (6.2039)
Am	AmN (5.005)	AmP (5.7114)	AmAs (5.876)	AmSb (6.240)	AmBi (6.332)
Sn		SnP (5.5359)	SnAs (5.716)	SnSb (6.130)	

compounds of the RE metals and that of the actinides crystallize in the NaCl structure, while those that contain other TMs prefer to crystallize in structures different from NaCl (NiAs or TiAs). Besides the values given in Tables 4.15–4.18, below we also list the lattice constants for alkali and some TM hydrides and also for TM borides and carbides:

Hydrides: LiH (4.0856 Å), NaH (4.880 Å), KH (5.704 Å), RbH (6.037 Å), CsH (6.376 Å), NiH (3.740 Å), and PdH (4.02 Å)

Borides: ZrB (4.65 Å), HfB (4.62 Å), and PuB (4.905 Å)

Carbides: ScC (4.51 Å), TiC (4.3186 Å), VC (4.182 Å), CrC (4.03 Å), ZrC (4.6828 Å), NbC (4.4691 Å), CeC (5.135 Å), HfC (4.63765 Å), TaC (4.4540 Å), ThC (5.346 Å), PaC (5.0608 Å), UC (4.9606 Å), NpC (5.005 Å), and PuC (4.731 Å).

Let us now proceed to calculate the limiting radius ratio for the NaCl structure. We can see in Fig. 4.26a to the right that 4 NNs of the Cl$^-$ ion, placed in the center of a cubic unit cell face, are located in the centers of the face edges. Figures 4.29–4.31 show the plane of one of the faces of the cube with cross sections of ions that, being

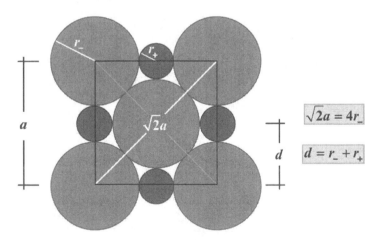

$$\sqrt{2}a = 4r_-$$

$$d = r_- + r_+$$

Figure 4.29 The plane of a face of the cubic F unit cell of the NaCl structure with the cross sections of 9 ions considered hard spheres. The large ion, located in the center of the face, makes contact with its NNs (small spheres) and also with the NNNs (large spheres). The NN distance d is equal to the sum of the ionic radii $r_- + r_+$.

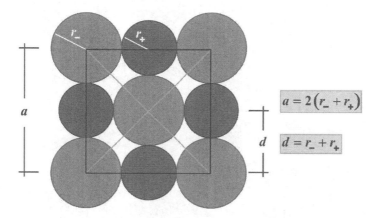

Figure 4.30 The same plane as in Fig. 4.29, but now the large ion, located in the center of the cube face, makes contact only with the NNs (small spheres). The NN distance, d, is equal to $r_- + r_+$.

considered hard spheres, are represented by circles on this plane. We can easily distinguish the following three cases:

(a) Each anion makes contact with its NNs (cations) and with the nearest anions as is shown in Fig. 4.29.
(b) Each anion makes contact only with its NNs (cations), see Fig. 4.30.
(c) Each anion makes contact only with the nearest anions as is illustrated in Fig. 4.31.

We will now proceed to calculate the r_+/r_- ratio for the case described in Fig. 4.29. We can see in this figure that

$$2r_- + 2r_+ = a, \tag{4.9}$$

as

$$\sqrt{2}a = 4r_-, \tag{4.10}$$

then

$$2r_- + 2r_+ = \frac{4r_-}{\sqrt{2}} = 2\sqrt{2}r_- \tag{4.11}$$

and finally

$$\frac{r_+}{r_-} = \sqrt{2} - 1 \cong 0.414. \tag{4.12}$$

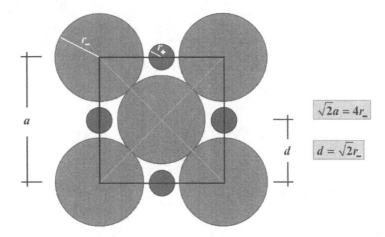

Figure 4.31 The same plane as in Figs. 4.29 and 4.30, but now the smaller ion is too small to make contact with larger ions and as a consequence the ion located in the center of the cube face makes contact only with its NNNs (large spheres). The NN distance, $d = \sqrt{2}r_-$, is defined only by the radius of the larger ion.

When $r_+/r_- = 0.414$, each anion touches both its NNs (cations) and the NNNs (anions). This is the limiting radius ratio for the NaCl structure which was already reported in Table 4.1. When the radius ratio is higher than the limiting one,

$$\frac{r_+}{r_-} > 0.414, \tag{4.13}$$

each large ion makes contact only with the NNs (small ions) but not with the NNNs (see Fig. 4.30) and the structure is stable. In the opposite case, when the radius ratio is smaller than the limiting one, each large ion is in contact only with the NNNs (see Fig. 4.31). However, in principle, this situation would lead to a less stable structure and in this case a lower coordination number is expected.

We can see in Table 4.19, in which are given the data for alkali halides, that the listed there radius ratios are smaller than the limiting one,

$$\frac{r_+}{r_-} < 0.414, \tag{4.14}$$

only for LiBr and LiI. In the rest of the alkali halides that have the NaCl structure, each ion touches its NNs that are of opposite sign.

Table 4.19 Several values for alkali halides: (a) cation and anion radii (to the right of the ion symbols), (b) ionic radius ratios (r_+/r_- or r_-/r_+), (c) sums of the ionic radii ($r_- + r_+$) in cases when $r_+/r_- > 0.414$ or $\sqrt{2}r_-$ in cases when $r_+/r_- < 0.414$, and (d) experimental values for the NN distances, $d = a/2$, where the lattice constants, a, are given in Table 4.15

	Li+ (0.76 Å)	Na+ (1.02 Å)	K+ (1.38 Å)	Rb+ (1.52 Å)	Cs+ (1.67 Å)
F− (1.33 Å)	r_+/r_-=0.57	r_+/r_-=0.77	r_-/r_+=0.96	r_-/r_+=0.88	r_-/r_+=0.80
	$r_- + r_+ = 2.09$	$r_- + r_+ = 2.35$	$r_- + r_+ = 2.71$	$r_- + r_+ = 2.85$	$r_- + r_+ = 3.00$
	$d = 2.01$	$d = 2.32$	$d = 2.67$	$d = 2.83$	$d = 3.01$
Cl− (1.81 Å)	r_+/r_-=0.42	r_+/r_-=0.56	r_+/r_-=0.76	r_+/r_-=0.84	
	$r_- + r_+ = 2.57$	$r_- + r_+ = 2.83$	$r_- + r_+ = 3.19$	$r_- + r_+ = 3.33$	
	$d = 2.56$	$d = 2.82$	$d = 3.15$	$d = 3.29$	
Br− (1.96 Å)	r_+/r_-=0.39	r_+/r_-=0.52	r_+/r_-=0.70	r_+/r_-=0.78	*cesium*
	$\sqrt{2}r_- = 2.77$	$r_- + r_+ = 2.98$	$r_- + r_+ = 3.34$	$r_- + r_+ = 3.48$	*chloride*
	$d = 2.75$	$d = 2.99$	$d = 3.30$	$d = 1.44$	*structures*
I− (2.20 Å)	r_+/r_-=0.35	r_+/r_-=0.46	r_+/r_-=0.63	r_+/r_-=0.69	
	$\sqrt{2}r_- = 3.11$	$r_- + r_+ = 3.22$	$r_- + r_+ = 3.58$	$r_- + r_+ = 3.72$	
	$d = 3.01$	$d = 3.24$	$d = 3.53$	$d = 3.67$	

We can also observe in this table that in the case of KF, RbF, and CsF compounds the cation radius is larger than the anion radius and, as a consequence, the condition

$$\frac{r_-}{r_+} > 0.414 \qquad (4.15)$$

has to be considered instead of the condition given by Eq. (4.13).

In Table 4.19 we list values for r_+/r_- or r_-/r_+ and $r_- + r_+$ or $\sqrt{2}r_-$, depending on the case in consideration. Those values were calculated using ionic radii given in the table. Table 4.19 contains also values for the distances, $d = a/2$, between the NNs, obtained using experimental lattice constants taken from Table 4.15. In those cases when the NNs touch each other, d should fulfill the equality

$$r_- + r_+ = d, \qquad (4.16)$$

what indeed happens to within 2% (see Table 4.19). This validates the concept of ionic radii, since the same radii can be used to calculate the interatomic distances for several compounds and those distances are very close to the experimental values obtained from the lattice constants.

In cases when the large ion makes contact only with its NNNs ($r_+/r_- < 0.414$), the distance to the NNs fulfills the following equality

$$\sqrt{2}r_- = d. \qquad (4.17)$$

This happens with very good accuracy for LiBr, and LiI (see Table 4.19).

Finally, we can observe in Table 4.19 that about half of the compounds considered there have their ionic radius ratios r_+/r_- (or r_-/r_+) in the range from 0.414 to 0.732, which is the expected range for the NaCl structure (see Table 4.1). The LiBr and LiI compounds represent the exceptions, for which the zinc blende (or wurtzite) structure is predicted according to the ranges for ionic radius ratios listed in Table 4.1. The other exceptions, NaF, KF, RbF, CsF, KCl, RbCl, and RbBr, have the radius ratios within the range corresponding to the CsCl structure.

In this section, we have learned that, among many other compounds, in the NaCl structure crystallize compounds of doubly ionized elements from columns II and VI of the periodic table, except for the beryllium compounds and MgTe. Geometric considerations, similar to that made for alkali halides, show that also in the case of II–VI compounds having the NaCl structure their ions may be considered, in good approximation, as hard impenetrable spheres of definite radii.

4.8 Cesium Chloride Structure

Pearson symbol: *cP*2, prototype: CsCl. In the NaCl structure (discussed in the previous section), the smaller in general cations are located in octahedral interstices (defined by 6 anions) present in the anionic *fcc* substructure. With the increase of the r_+/r_- ratio, a cubic interstice defined by 8 anions becomes a better option for the cations. This is the case of the cesium chloride (CsCl) structure, for which the limiting radius ratio is 0.732. The CsCl structure is a superposition of two simple cubic substructures. Both, the cations and the anions, occupy the cubic interstices present in each substructure. In Fig. 4.32, we show the smallest unit cell for CsCl. This cell is defined by anions and the cation is in its center.

One of the two principal groups of compounds that, under normal conditions, crystallize in the CsCl structure is formed by three cesium halides: CsBr, CsCl, and CsI, and also three thallium halides: TlBr, TlCl, and TlI. We can see in Table 4.19 that CsBr, CsCl, and CsI are the halides of the largest univalent ions (remember

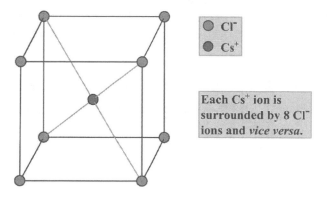

● Cl⁻
● Cs⁺

Each Cs⁺ ion is surrounded by 8 Cl⁻ ions and *vice versa.*

Figure 4.32 A conventional unit cell for cesium chloride with two ions in it.

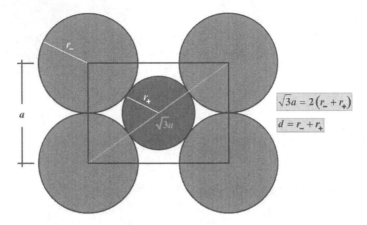

$$\sqrt{3}a = 2\left(r_- + r_+\right)$$

$$d = r_- + r_+$$

Figure 4.33 A plane defined by two body diagonals of the cube shown in Fig. 4.32. In this plane, there are the points of contact between the cation and its four NNs.

that these ions have somewhat different radii in the case of the coordination number 8). The other (quite numerous) group of compounds that crystallize in the CsCl structure is formed by intermetallic compounds.

Let us now calculate the limiting radius ratio for the CsCl structure. As in the case of the *bcc* structure, the ions that are at the vertices of the cube are the NNs of the ion that is in the center of the cube. Figure 4.33 shows a plane defined by two body diagonals of the cube with the cross section of a cation placed in the center and the cross sections of four anions placed at the vertices of the cube. We can see in the figure that the points of contact between the cation and the anions are on the body diagonals of the cube, therefore,

$$2r_- + 2r_+ = \sqrt{3}a. \qquad (4.18)$$

Since $a = 2r_-$ then

$$2r_- + 2r_+ = 2\sqrt{3}r_-, \qquad (4.19)$$

and finally

$$\frac{r_+}{r_-} = \sqrt{3} - 1 \cong 0.732. \qquad (4.20)$$

Equation (4.20) gives the value for the limiting radius ratio for the CsCl structure. This value was already included in Table 4.1. In that

Table 4.20 Lattice constants, obtained under normal conditions, of cesium and thallium halides that crystallize in the CsCl structure

Compound	$a(\text{Å})$	Compound	$a(\text{Å})$
CsBr	4.286	TlBr	3.970
CsCl	4.123	TlCl	3.834
CsI	4.567	TlI	4.205

limiting case each anion touches both its NNs (cations) and the NNNs (anions).

In Tables 4.20–4.23 we list the data for about 200 compounds that crystallize in the CsCl structure. Table 4.20 gives the lattice constants for cesium and thallium halides, while Tables 4.21–4.23 report the data for intermetallic compounds. In those intermetallic compounds that are listed in Tables 4.21 and 4.22 one of the metallic elements is a RE metal. We can observe in these tables

Table 4.21 Lattice constants (in Angstroms) obtained under normal conditions for intermetallic compounds of the RE-Mg or RE-III type that crystallize in the CsCl structure. The elements in the compound symbols are listed alphabetically

	Mg	Al	In	Tl
Sc	MgSc (3.597)	AlSc (3.450)		
Y	MgY (3.79)	AlY (3.754)	InY (3.806)	TlY (3.751)
La	LaMg (3.965)		InLa (3.985)	LaTl (3.922)
Ce	CeMg (3.899)	AlCe (3.86)		CeTl (3.893)
Pr	MgPr (3.888)	AlPr (3.82)	InPr (3.955)	PrTl (3.869)
Nd	MgNd (3.867)	AlNd (3.73)		NdTl (3.848)
Sm	MgSm (3.848)	AlSm (3.739)	InSm (3.815)	SmTl (3.813)
Eu				EuTl (3.975)
Gd	GdMg (3.824)	AlGd (3.7208)	GdIn (3.830)	GdTl (3.7797)
Tb	MgTb (3.784)			TbTl (3.760)
Dy	DyMg (3.776)	AlDy (3.6826)	DyIn (3.7866)	DyTl (3.743)
Ho	HoMg (3.770)		HoIn (3.774)	HoTl (3.735)
Er	ErMg (3.758)		ErIn (3.745)	ErTl (3.715)
Tm	MgTm (3.744)		InTm (3.737)	TlTm (3.711)
Yb			InYb (3.8138)	TlYb (3.826)
Lu	LuMg (3.727)			

Table 4.22 Lattice constants (in Angstroms) obtained under normal conditions for intermetallic compounds of the RE-TM type that crystallize in the CsCl structure. The elements in the compound symbols are listed alphabetically

	Cu	Zn	Rh	Ag	Cd	Au	Hg
Sc	CuSc 3.256	ScZn 3.35	RhSc 3.206	AgSc 3.412	CdSc 3.513	AuSc 3.370	HgSc 3.480
Y	CuY 3.4757	YZn 3.577	RhY 3.410	AgY 3.6196	CdY 3.719	AuY 3.559	HgY 3.682
La		LaZn 3.759		AgLa 3.814	CdLa 3.904		HgLa 3.845
Ce		CeZn 3.696		AgCe 3.755	CdCe 3.855		CeHg 3.815
Pr		PrZn 3.678		AgPr 3.746	CdPr 3.829	AuPr 3.68	HgPr 3.799
Nd		NdZn 3.667		AgNd 3.716	CdNd 3.819	AuNd 3.659	HgNd 3.780
Sm	CuSm 3.528	SmZn 3.627	RhSm 3.466	AgSm 3.673	CdSm 3.779	AuSm 3.621	HgSm 3.744
Eu	CuEu 3.479	EuZn 3.808			CdEu 3.951		EuHg 3.880
Gd	CuGd 3.501	GdZn 3.609	GdRh 3.435	AgGd 3.6491	CdGd 3.748	AuGd 3.6009	GdHg 3.719
Tb	CuTb 3.480	TbZn 3.576	RhTb 3.417	AgTb 3.627	CdTb 3.723	AuTb 3.576	HgTb 3.678
Dy	CuDy 3.462	DyZn 3.562	DyRh 3.403	AgDy 3.609	CdDy 3.716	AuDy 3.555	DyHg 3.676
Ho	CuHo 3.445	HoZn 3.548	HoRh 3.377	AgHo 3.601	CdHo 3.701	AuHo 3.541	HgHo 3.660
Er	CuEr 3.430	ErZn 3.532	ErRh 3.361	AgEr 3.574	CdEr 3.685	AuEr 3.5346	ErHg 3.645
Tm	CuTm 3.414	TmZn 3.516	RhTm 3.358	AgTm 3.562	CdTm 3.663	AuTm 3.516	HgTm 3.632
Yb		YbZn 3.629	RhYb 3.347	AgYb 3.6787	CdYb 3.8086	AuYb 3.5634	HgYb 3.735
Lu		LuZn 3.491	LuRh 3.334		CdLu 3.640	AuLu 3.4955	HgLu 3.607

Table 4.23 Lattice constants obtained under normal conditions for intermetallic compounds that crystallize in the CsCl structure. The elements are listed alphabetically in those compounds where at least one of the elements is a TM

Compound	$a(Å)$	Compound	$a(Å)$	Compound	$a(Å)$
β-AgCd	3.332	CaTl	3.851	HoIr*	3.383
AgGa	3.171	CdSr	4.003	InNi	3.093
β-AgLi	3.168	CoFe	2.857	InPd	3.246
β-AgMg	3.124	CoGa	2.880	IrLu*	3.332
β-AgZn	3.1558	CoHf	3.164	IrSc*	3.205
β-AlCo	2.864	CoSc*	3.145	LiTl	3.435
AlFe	2.908	CoTi	2.995	LuPd*	3.415
AlIr	2.983	CoZr	3.181	MgRh	3.099
β-AlNi	2.882	β-CuPd	2.988	MgSr	3.908
AlOs	3.001	β-CuZn	2.950	MgTl	3.635
AuCd	3.3232	CuZr	3.2620	β-MnRh	3.044
AuCs	4.262	FeRh	2.983	NiSc*	3.171
AuMg	3.266	FeTi	2.976	NiTi	3.01
β-AuZn	3.1485	FeV	2.910	OsTi	3.07
BaCd	4.207	GaIr	3.004	PdSc*	3.282
BaHg	4.125	β-GaNi	2.886	PtSc*	3.268
BeCo	2.624	GaRh	3.0063	RuSc*	3.203
BeCu	2.702	GaRu	3.010	RuTi	3.06
BeNi	2.6121	HgLi	3.287	SrTl	4.038
BePd	2.813	HgMg	3.448	TlBi	3.98
BeRh	2.740	HgMn	3.316	ZnZr	3.336
CaHg	3.759	HgSr	3.930		

*Intermetallic binary compounds where one of the elements is a RE metal

that the number of such compounds is significant. In the case of the intermetallic compounds in which at least one of the elements is a TM, we adopted the convention according to which the elements in compound symbols are listed alphabetically.

Problems

Exercise 4.1 How many cations and anions do belong to the cubic F unit cell of the zinc blende (β-ZnS) structure? Draw this cell and the ions belonging to it. Show the coordinate triplets of each ion expressing its coordinates in units of the lattice constant a.

Exercise 4.2 Repeat Exercise 4.1, but now for the fluorite (CaF_2) structure.

Exercise 4.3 Table 4.10 lists the distances of an ion to the NNs in MnSe, MnTe, ZnSe, CdSe, and GaN, for two crystal structures: zinc blende and wurtzite. These distances have been obtained using experimental lattice constants. Make a similar table with the distances d_{NNN} of an ion to the NNNs. Express, in percentage, the difference between d_{NNN} obtained for the zinc blende and wurtzite type structures. In your calculations use the experimental lattice constants listed in Tables 4.3, 4.4, and 4.9.

Exercise 4.4

(a) Draw a hexagonal prism for β-ZnS, which crystallizes in the zinc blende structure. This prism should be able to reproduce the β-ZnS structure. Show on the figure two-dimensional *hcp* layers A_{Zn}, B_{Zn}, C_{Zn}, A_S, B_S, and C_S. Express the distances between the consecutive layers in terms of the hexagonal prism heights c.

(b) How many ions of each type do belong to the hexagonal prism you have drawn in (a) and how many ions do belong to the hexagonal prism for the wurtzite structure?

Exercise 4.5 The sets of ions shown in Figs. 4.10a and 4.10b have threefold rotation axes, however the highest symmetry axis of an infinite wurtzite structure is a sixfold screw axis 6_3. Show such axis in the infinite volume of the wurtzite structure.

Exercise 4.6 Draw the graphical symbols of the highest-order symmetry axes for the sets of ions shown in Fig. 4.18a to the left and to the right.

Exercise 4.7 The space group of the NiAs structure is the same as the common space group of the following structures discussed in Chapter 3: *hcp*, *dhcp*, and graphite. Its full Hermann–Mauguin symbol is $P6_3/m2/m2/c$. Show the sixfold screw axis with center of symmetry, $6_3/m$, in the infinite volume of the NiAs structure.

Exercise 4.8 Let us consider 8 and 12 closest cations to a given cation in the NiAs and anti-NiAs structures, respectively. In the case of the NiAs structure the 8 cations can be divided, according to the distance to the cation in consideration, into two groups of $n_{cc}^I = 2$ and $n_{cc}^{II} = 6$ ions which are closer and more distant to the cation, respectively. Using similar criterion, the 12 cations considered in the anti-NiAs structure can be divided into two groups of 6 ions each. In each case the distances depend on the lattice constant ratio c/a.

(a) For the following compounds: VSb ($c/a = 1.28$), VSe ($c/a = 1.63$), VS ($c/a = 1.75$), and TiS ($c/a = 1.93$), that crystallize in the NiAs structure, calculate the two closest cation-cation distances, d_{cc}^I and d_{cc}^{II}. For each compound, compare the obtained distances expressing the difference in percentage. How does this difference change with the increase of the c/a ratio? How many NNs and NNNs does a cation have in each compound? Use the lattice constants a and c from Table 4.13 for VSb and from Table 4.14 for TiS, VS, and VSe.

(b) Do a similar work as in (a) for the anti-NiAs structure on the example of the following compounds: PtB ($c/a = 1.21$), ScPo ($c/a = 1.65$), δ'-NbN ($c/a = 1.87$), and VP ($c/a = 1.96$). How many NNNs does each Pt ion have in the PtB compound (and also V ion in the VP compound)? To which two-dimensional *hcp* layer do those ions belong? Answer similar questions for the ScPo and δ'-NbN compounds. Use lattice constants a and c from Table 4.11 for PtB, from Table 4.13 for δ'-NbN and VP, and from Table 4.14 for ScPo.

Exercise 4.9 In the case when the lattice constant ratio c/a is much smaller than 1.633, each cation in the NiAs structure has indeed 8 NNs: 6 anions and 2 cations.

(a) Show that the above is true, to within 3% of the NN interatomic distance, for the following compounds: CuSb ($c/a = 1.34$), PdSb ($c/a = 1.37$), IrSb ($c/a = 1.39$), and IrTe ($c/a = 1.37$). Use lattice constants a and c from Table 4.13 for CuSb, PdSb, and IrSb and from Table 4.14 for IrTe.

(b) Compare the cation-cation distances d_{cc} calculated in (a) with the distances $d_{XX}^{element}$ (X = Cu, Pd, or Ir) between Cu, Pd, and Ir NNs in the crystals of these elements. Confirm that in each case d_{cc} differ from $d_{XX}^{element}$ by less than 2%. Use the NN interatomic distances for elements, listed in Table 3.16 in Section 3.17.

Exercise 4.10 Tables 4.13 and 4.18 report the experimental data obtained under normal conditions for ZrP in the β (of TiAs type) and α (of NaCl type) phases, respectively.

(a) Calculate and compare the distances d_{ac}^{α} and d_{ac}^{β} between the NNs in the two structures.

(b) Calculate the distance d_{aa}^{α} of the NNNs of an anion in α-ZrP. Calculate also the distance d_{aa}^{β}, between a given anion P and the 12 closest anions to it in β-ZrP. Compare the obtained distances expressing the difference in percentage.

(c) Calculate the distance d_{cc}^{α} of the NNNs of a cation in α-ZrP. Calculate also the distances d_{cc}^{β} between a given cation Zr and the 10 closest Zr cations to it in β-ZrP. Compare the obtained distances expressing the differences in percentage.

Exercise 4.11 Tables 4.13 and 4.18 report the experimental data obtained under ambient conditions for δ'-NbN (anti-NiAs structure) and ε-NbN (TiAs structure), and δ-NbN (NaCl structure), respectively.

(a) Calculate and compare the NN distances d_{ac} for the three structures.

(b) Calculate the distance d_{aa} of the NNNs of an anion in the three structures. Compare the distances obtained for δ'-NbN and ε-NbN with that obtained for δ-NbN.

(c) Do the same as in (b) for NNNs of a cation.

Exercise 4.12 Find the filling factors for the following compounds that crystallize in the NaCl structure: LiCl ($r_+/r_- = 0.42$), NaCl

$(r_+/r_- = 0.56)$, RbCl $(r_+/r_- = 0.84)$, and KF $(r_-/r_+ = 0.96)$. The cation radii, r_+, and the anion radii, r_-, are listed in Table 4.19 and the lattice parameters a for I-VII compounds are listed in Table 4.15.

(a) What is the relation between the values for the filling factor and the r_+/r_- (or r_-/r_+) ratio?
(b) What would be the value of the filling factor in the case when $r_+/r_- = 1$? Answer without doing any calculations.

Exercise 4.13 Prepare a similar table to Table 4.19 for II-VI compounds that crystallize in the NaCl structure. The radii of the double ionized elements from column II of the periodic table are: 0.72 Å for Mg^{2+}, 1.00 Å for Ca^{2+}, 1.18 Å for Sr^{2+}, and 1.35 Å for Ba^{2+}, and the radii for the double ionized elements from column VI are: 1.40 Å for O^{2-}, 1.84 Å for S^{2-}, 1.98 Å for Se^{2-}, and 2.21 Å for Te^{2-}. Show that the interatomic distance d expressed by the sum of ionic radii $r_- + r_+$, if $r_+/r_- > 0.414$, or by $\sqrt{2}r_-$, if $r_+/r_- < 0.414$, agree to within 2% with the value obtained from the experimental lattice constant $(d = a/2)$. The lattice constants a for II-VI compounds are listed in Table 4.16.

Exercise 4.14 Draw a second choice, different from that shown in Fig. 4.32, for the conventional unit cell of cesium chloride.

Chapter 5

Reciprocal Lattice

5.1 Introduction

Crystal structures considered in previous chapters correspond to ideal crystalline materials and refer to the cases where the atoms are in their equilibrium positions, which obviously represents the first-order approximation in the description of such materials. We know already that an infinite crystal structure possesses a translation symmetry which is at the same time the translation symmetry of its lattice. In this chapter, we will introduce the concept of the so-called *reciprocal lattice*, which has the same point-group symmetry as the crystal lattice (direct lattice) and plays an important role in the description of the physical properties of crystalline materials.

5.2 The Concept of the Reciprocal Lattice

The concept of the reciprocal lattice will be introduced starting from the fact that in an ideal infinite crystalline material the electrostatic potential produced by all the charges present in it is periodic with the periodicity of the crystal lattice. Let us denote the lattice translation vector as $\vec{a}_{\vec{n}}$. This vector can be expressed as a linear

Basic Elements of Crystallography (2nd Edition)
Nevill Gonzalez Szwacki and Teresa Szwacka
Copyright © 2016 Pan Stanford Publishing Pte. Ltd.
ISBN 978-981-4613-57-6 (Hardcover), 978-981-4613-58-3 (eBook)
www.panstanford.com

combination of three non-collinear and not all in the same plane primitive translation vectors $\vec{a}_1, \vec{a}_2, \vec{a}_3$ or as a linear combination of the unit vectors $\hat{a}_1, \hat{a}_2, \hat{a}_3$

$$\vec{a}_{\vec{n}} = n_1\vec{a}_1 + n_2\vec{a}_2 + n_3\vec{a}_3 = n_1a_1\hat{a}_1 + n_2a_2\hat{a}_2 + n_3a_3\hat{a}_3, \qquad (5.1)$$

where n_1, n_2, and $n_3 \in \mathbb{Z}$. A position vector $\vec{r}$ of any point in the crystal may be expressed in terms of the $\hat{a}_1, \hat{a}_2, \hat{a}_3$ basis as

$$\vec{r} = \xi_1\hat{a}_1 + \xi_2\hat{a}_2 + \xi_3\hat{a}_3, \qquad (5.2)$$

where the real numbers ξ_1, ξ_2, and ξ_3 are components of the vector in this basis.

In the approximation in which we are considering the crystalline material, the points at $\vec{r}$ and $\left(\vec{r} + \vec{a}_{\vec{n}}\right)$, shown in Fig. 5.1, are phys-ically equivalent and as a consequence the electrostatic potential, $V(\vec{r})$, produced by all the charges present in the crystal has the same value in both points:

$$V(\vec{r}) = V\left(\vec{r} + \vec{a}_{\vec{n}}\right). \qquad (5.3)$$

That means the potential is periodic. Any periodic function can be expanded into its Fourier series. We will do that for $V(\vec{r})$ with respect to each of the components of the argument $\vec{r}$ in axes ξ_1, ξ_2,

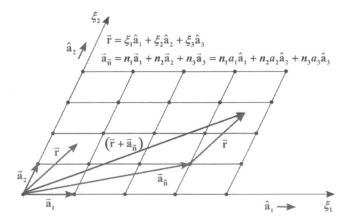

Figure 5.1 A two-dimensional crystal lattice. The points at $\vec{r}$ and $(\vec{r} + \vec{a}_{\vec{n}})$ have equivalent positions in the lattice.

and ξ_3, along which the periodicity occurs. We have then

$$V(\vec{r}) = V(\xi_1, \xi_2, \xi_3) = \sum_{l_1=-\infty}^{\infty} \sum_{l_2=-\infty}^{\infty} \sum_{l_3=-\infty}^{\infty} V_{l_1 l_2 l_3}$$

$$\times \exp\left[2\pi i \left(\frac{l_1 \xi_1}{a_1} + \frac{l_2 \xi_2}{a_2} + \frac{l_3 \xi_3}{a_3}\right)\right], \qquad (5.4)$$

where l_1, l_2, and l_3 are integer numbers and $V(\xi_1, \xi_2, \xi_3)$ is periodic with respect to each of its arguments ξ_1, ξ_2, and ξ_3 with periods a_1, a_2, and a_3, respectively. It is easy to show that the potential expressed in this way is indeed periodic. Since

$$\vec{r} + \vec{a}_{\vec{n}} = (\xi_1 + n_1 a_1)\,\hat{a}_1 + (\xi_2 + n_2 a_2)\,\hat{a}_2 + (\xi_3 + n_3 a_3)\,\hat{a}_3, \qquad (5.5)$$

we have

$$V\left(\vec{r} + \vec{a}_{\vec{n}}\right) = \sum_{l_1=-\infty}^{\infty} \sum_{l_2=-\infty}^{\infty} \sum_{l_3=-\infty}^{\infty} V_{l_1 l_2 l_3} \exp\left\{2\pi i \left[\frac{l_1\,(\xi_1 + n_1 a_1)}{a_1}\right.\right.$$

$$\left.\left. + \frac{l_2\,(\xi_2 + n_2 a_2)}{a_2} + \frac{l_3\,(\xi_3 + n_3 a_3)}{a_3}\right]\right\}$$

$$= \sum_{l_1=-\infty}^{\infty} \sum_{l_2=-\infty}^{\infty} \sum_{l_3=-\infty}^{\infty} V_{l_1 l_2 l_3} \exp[2\pi i\,(l_1 n_1 + l_2 n_2 + l_3 n_3)]$$

$$\times \exp\left[2\pi i \left(\frac{l_1 \xi_1}{a_1} + \frac{l_2 \xi_2}{a_2} + \frac{l_3 \xi_3}{a_3}\right)\right]$$

$$= \sum_{l_1=-\infty}^{\infty} \sum_{l_2=-\infty}^{\infty} \sum_{l_3=-\infty}^{\infty} V_{l_1 l_2 l_3}$$

$$\times \exp\left[2\pi i \left(\frac{l_1 \xi_1}{a_1} + \frac{l_2 \xi_2}{a_2} + \frac{l_3 \xi_3}{a_3}\right)\right]$$

$$= V(\vec{r}), \qquad (5.6)$$

where we took into account that

$$l_1 n_1 + l_2 n_2 + l_3 n_3 = \text{(integer number)}$$

$$\Rightarrow \quad \exp\left[2\pi i\,(l_1 n_1 + l_2 n_2 + l_3 n_3)\right] = 1 \quad (5.7)$$

Now, we will make a transformation to an orthogonal coordinate system. The components of the position vector $\vec{r}$ in the orthogonal system shown in Fig. 5.2 can be expressed as a function of its components given in the ξ_1 and ξ_2 axes:

$$\begin{cases} x_1 = \xi_1 + \xi_2 \cos\varphi \\ x_2 = \xi_2 \sin\varphi \end{cases}. \qquad (5.8)$$

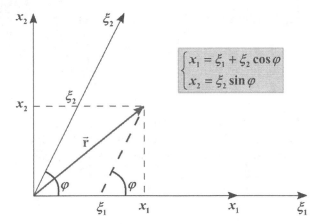

Figure 5.2 The relation between the components of the vector $\vec{r}$ in the orthogonal and non-orthogonal coordinate systems.

The above equations represent a system of linear equations for ξ_1 and ξ_2. By solving for ξ_2 first and ξ_1 after, we find

$$\begin{cases} \xi_2 = \dfrac{1}{\sin\varphi}x_2 \\ \xi_1 = x_1 - \dfrac{\cos\varphi}{\sin\varphi}x_2 \end{cases} \Rightarrow \begin{cases} \xi_1 = x_1 - x_2\cot\varphi \\ \xi_2 = x_2\csc\varphi \end{cases}. \qquad (5.9)$$

The formulas for ξ_1 and ξ_2 can be rewritten more generally as follows:

$$\begin{cases} \xi_1 = a_{11}x_1 + a_{12}x_2, & \text{where} \quad a_{11} = 1 \text{ and } a_{12} = -\cot\varphi \\ \xi_2 = a_{21}x_1 + a_{22}x_2, & \text{where} \quad a_{21} = 0 \text{ and } a_{22} = \csc\varphi \end{cases}. \qquad (5.10)$$

In a three-dimensional case, if the origins of the non-orthogonal and orthogonal coordinate systems coincide, we have

$$\begin{cases} \xi_1 = a_{11}x_1 + a_{12}x_2 + a_{13}x_3 = \sum\limits_{k=1}^{3} a_{1k}x_k \\ \xi_2 = a_{21}x_1 + a_{22}x_2 + a_{23}x_3 = \sum\limits_{k=1}^{3} a_{2k}x_k \, , \\ \xi_3 = a_{31}x_1 + a_{32}x_2 + a_{33}x_3 = \sum\limits_{k=1}^{3} a_{3k}x_k \end{cases} \qquad (5.11)$$

where the a_{ik} coefficients are determined by the angles between axes ξ_i and x_k.

We will now substitute ξ_1, ξ_2, and ξ_3 given by Eqs. 5.11 into Eq. 5.4, so that

$$V(\vec{r}) = \sum_{l_1=-\infty}^{\infty} \sum_{l_2=-\infty}^{\infty} \sum_{l_3=-\infty}^{\infty} V_{l_1 l_2 l_3} \exp\left[2\pi i A\right], \qquad (5.12)$$

where

$$
\begin{aligned}
A &= \frac{l_1}{a_1} \sum_{k=1}^{3} a_{1k} x_k + \frac{l_2}{a_2} \sum_{k=1}^{3} a_{2k} x_k + \frac{l_3}{a_3} \sum_{k=1}^{3} a_{3k} x_k \\
&= \left(\frac{l_1 a_{11}}{a_1} + \frac{l_2 a_{21}}{a_2} + \frac{l_3 a_{31}}{a_3} \right) x_1 + \left(\frac{l_1 a_{12}}{a_1} + \frac{l_2 a_{22}}{a_2} + \frac{l_3 a_{32}}{a_3} \right) x_2 \\
&\quad + \left(\frac{l_1 a_{13}}{a_1} + \frac{l_2 a_{23}}{a_2} + \frac{l_3 a_{33}}{a_3} \right) x_3 \qquad (5.13)
\end{aligned}
$$

and abbreviating

$$
\begin{cases}
b_1 = 2\pi \left(\dfrac{l_1 a_{11}}{a_1} + \dfrac{l_2 a_{21}}{a_2} + \dfrac{l_3 a_{31}}{a_3} \right) \\[2mm]
b_2 = 2\pi \left(\dfrac{l_1 a_{12}}{a_1} + \dfrac{l_2 a_{22}}{a_2} + \dfrac{l_3 a_{32}}{a_3} \right) \\[2mm]
b_3 = 2\pi \left(\dfrac{l_1 a_{13}}{a_1} + \dfrac{l_2 a_{23}}{a_2} + \dfrac{l_3 a_{33}}{a_3} \right)
\end{cases}
\qquad (5.14)
$$

we obtain

$$V(\vec{r}) = \sum_{l_1=-\infty}^{\infty} \sum_{l_2=-\infty}^{\infty} \sum_{l_3=-\infty}^{\infty} V_{l_1 l_2 l_3} \exp\left[i \left(b_1 x_1 + b_2 x_2 + b_3 x_3\right)\right].$$

$$(5.15)$$

From this point on, the summation over l_1, l_2, and l_3 will be replaced by the summation over discreet parameters b_1, b_2, and b_3, determined by the l_k ($k = 1, 2, 3$) according to Eqs. 5.14. Moreover, it will be helpful to consider b_i ($i = 1, 2, 3$) as coordinates of a certain vector $\vec{b}$ in the orthogonal coordinate system. In this manner

$$b_1 x_1 + b_2 x_2 + b_3 x_3 = \vec{b} \cdot \vec{r} \qquad (5.16)$$

and then

$$
\begin{aligned}
V(\vec{r}) &= \sum_{b_1} \sum_{b_2} \sum_{b_3} V_{b_1 b_2 b_3} \exp\left[i \left(b_1 x_1 + b_2 x_2 + b_3 x_3\right)\right] \\
&= \sum_{\vec{b}} V_{\vec{b}} \exp\left(i \vec{b} \cdot \vec{r}\right). \qquad (5.17)
\end{aligned}
$$

The components of vector $\vec{b}$ are given by Eqs. 5.14; however, it is convenient to determine the formula for this vector again, starting from the condition of periodicity of the crystal potential, which guides us to the following conclusion:

$$
\left.
\begin{aligned}
V\left(\vec{r}+\vec{a}_{\vec{n}}\right) &= \sum_{\vec{b}} V_{\vec{b}}\exp\left[i\vec{b}\cdot\left(\vec{r}+\vec{a}_{\vec{n}}\right)\right] \\
&= \sum_{\vec{b}} V_{\vec{b}}\exp\left(i\vec{b}\cdot\vec{r}\right)\cdot\exp\left(i\vec{b}\cdot\vec{a}_{\vec{n}}\right) \\
&= V(\vec{r}) \\
\text{and} \\
V(\vec{r}) &= \sum_{\vec{b}} V_{\vec{b}}\exp\left(i\vec{b}\cdot\vec{r}\right)
\end{aligned}
\right\} \Rightarrow \exp\left(i\vec{b}\cdot\vec{a}_{\vec{n}}\right)=1.
$$

$$(5.18)$$

This means that in order for the potential to be periodic with periods $\vec{a}_{\vec{n}}$, the following equality has to be achieved:

$$\exp\left(i\vec{b}\cdot\vec{a}_{\vec{n}}\right) = 1. \tag{5.19}$$

It is easy to see that Eq. 5.19 implies the periodicity of the function $\exp\left(i\vec{b}\cdot\vec{r}\right)$, since

$$\exp\left[i\vec{b}\cdot\left(\vec{r}+\vec{a}_{\vec{n}}\right)\right] = \exp\left(i\vec{b}\cdot\vec{r}\right)\cdot\exp\left(i\vec{b}\cdot\vec{a}_{\vec{n}}\right) = \exp\left(i\vec{b}\cdot\vec{r}\right)$$

$$\Downarrow$$

$$\exp\left[i\vec{b}\cdot\left(\vec{r}+\vec{a}_{\vec{n}}\right)\right] = \exp\left(i\vec{b}\cdot\vec{r}\right) \tag{5.20}$$

and *vice versa*; Eq. 5.20 implies Eq. 5.19. In conclusion, the potential $V(\vec{r})$ can be expressed as a function of plane waves $\exp\left(i\vec{b}\cdot\vec{r}\right)$, which are periodic with the periodicity of the lattice. Next, we will use Eq. 5.19 to find the expression for the vector $\vec{b}$ that characterizes such plane waves. We have

$$\exp\left(i\vec{b}\cdot\vec{a}_{\vec{n}}\right) = 1 \Rightarrow \vec{b}\cdot\vec{a}_{\vec{n}} = \text{(integer number)}\cdot 2\pi, \text{ for all vectors } \vec{n} \tag{5.21}$$

and using Eq. 5.1 we obtain

$$
\begin{aligned}
\vec{b}\cdot\vec{a}_{\vec{n}} &= \vec{b}\cdot(n_1\vec{a}_1 + n_2\vec{a}_2 + n_3\vec{a}_3) \\
&= n_1\left(\vec{b}\cdot\vec{a}_1\right) + n_2\left(\vec{b}\cdot\vec{a}_2\right) + n_3\left(\vec{b}\cdot\vec{a}_3\right) \\
&= \text{(integer number)}\cdot 2\pi,
\end{aligned}
\tag{5.22}
$$

for all possible n_1, n_2, and $n_3 \in \mathbb{Z}$. The condition given by Eq. 5.22 is satisfied only if

$$\begin{cases} \vec{b} \cdot \vec{a}_1 = 2\pi \cdot g_1 \\ \vec{b} \cdot \vec{a}_2 = 2\pi \cdot g_2 \text{ , where } g_1, g_2, \text{ and } g_3 \in \mathbb{Z}. \\ \vec{b} \cdot \vec{a}_3 = 2\pi \cdot g_3 \end{cases} \qquad (5.23)$$

The above represents three scalar equations for three components of vector $\vec{b}$. To solve these equations, instead of using the orthogonal coordinate system, we are going to express vector $\vec{b}$ as a linear combination of three non-collinear and not all in the same plane vectors defined in the following manner:

$$\vec{a}_1 \times \vec{a}_2, \ \vec{a}_2 \times \vec{a}_3, \ \vec{a}_3 \times \vec{a}_1.$$

We have then

$$\vec{b} = \alpha \, (\vec{a}_1 \times \vec{a}_2) + \beta \, (\vec{a}_2 \times \vec{a}_3) + \gamma \, (\vec{a}_3 \times \vec{a}_1), \qquad (5.24)$$

where the scalars α, β, and γ are coefficients of the linear combination. The task of solving Eqs. 5.23 consists now in finding the expression for the α, β, and γ coefficients. To this end we substitute Eq. 5.24 into Eqs. 5.23 and obtain

$$\begin{cases} \vec{b} \cdot \vec{a}_1 = \beta \, (\vec{a}_2 \times \vec{a}_3) \cdot \vec{a}_1 = 2\pi \cdot g_1 \\ \vec{b} \cdot \vec{a}_2 = \gamma \, (\vec{a}_3 \times \vec{a}_1) \cdot \vec{a}_2 = 2\pi \cdot g_2 \implies \\ \vec{b} \cdot \vec{a}_3 = \alpha \, (\vec{a}_1 \times \vec{a}_2) \cdot \vec{a}_3 = 2\pi \cdot g_3 \end{cases} \begin{cases} \beta \Omega_0 = 2\pi \cdot g_1 \\ \gamma \Omega_0 = 2\pi \cdot g_2 \text{ ,} \\ \alpha \Omega_0 = 2\pi \cdot g_3 \end{cases}$$
$$(5.25)$$

since

$$(\vec{a}_1 \times \vec{a}_2) \cdot \vec{a}_3 = (\vec{a}_2 \times \vec{a}_3) \cdot \vec{a}_1 = (\vec{a}_3 \times \vec{a}_1) \cdot \vec{a}_2 = \Omega_0. \qquad (5.26)$$

From Eqs. 5.25, we finally obtain the following expressions for the coefficients α, β, and γ:

$$\alpha = 2\pi \frac{g_3}{\Omega_0}, \ \beta = 2\pi \frac{g_1}{\Omega_0}, \ \gamma = 2\pi \frac{g_2}{\Omega_0}. \qquad (5.27)$$

Then vector $\vec{b}$ has the following formula:

$$\begin{aligned} \vec{b} &= g_3 2\pi \frac{(\vec{a}_1 \times \vec{a}_2)}{\Omega_0} + g_1 2\pi \frac{(\vec{a}_2 \times \vec{a}_3)}{\Omega_0} + g_2 2\pi \frac{(\vec{a}_3 \times \vec{a}_1)}{\Omega_0} \\ &= g_1 \vec{b}_1 + g_2 \vec{b}_2 + g_3 \vec{b}_3, \end{aligned} \qquad (5.28)$$

where

$$\vec{b}_1 = 2\pi \frac{(\vec{a}_2 \times \vec{a}_3)}{\Omega_0}, \ \vec{b}_2 = 2\pi \frac{(\vec{a}_3 \times \vec{a}_1)}{\Omega_0}, \ \vec{b}_3 = 2\pi \frac{(\vec{a}_1 \times \vec{a}_2)}{\Omega_0}. \qquad (5.29)$$

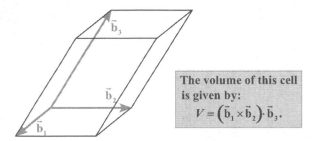

Figure 5.3 The unit cell of a reciprocal lattice defined by the primitive translation vectors $\vec{b}_1$, $\vec{b}_2$, and $\vec{b}_3$.

We obtained that vector $\vec{b}$ is a linear combination of vectors $\vec{b}_1$, $\vec{b}_2$, $\vec{b}_3$, defined by Eqs. 5.29, with integer coefficients g_1, g_2, and g_3. We have then a set of discreet vectors $\vec{b}$ and this makes them similar to vectors $\vec{a}_{\vec{n}}$, defined by Eq. 5.1, that go from one point to any other of the crystal lattice (direct lattice). In analogy to $\vec{a}_{\vec{n}}$ we define vectors $\vec{b}_{\vec{g}}$ as

$$\vec{b}_{\vec{g}} = g_1\vec{b}_1 + g_2\vec{b}_2 + g_3\vec{b}_3,\ g_1, g_2,\ \text{and}\ g_3 \in \mathbb{Z}. \qquad (5.30)$$

It is convenient to call "lattice" a set of points generated by all possible vectors $\vec{b}_{\vec{g}}$. This lattice is called the *reciprocal lattice* and the vectors $\vec{b}_1$, $\vec{b}_2$, and $\vec{b}_3$ are its primitive translation vectors. From Eqs. 5.29, we can see that they are defined by vectors $\vec{a}_1$, $\vec{a}_2$, $\vec{a}_3$ that are three non-collinear and not all in the same plane primitive translation vectors of the direct lattice. The primitive translation vectors of the reciprocal lattice define the unit cell of this lattice, which is shown in Fig. 5.3.

In conclusion, we can say that the plane waves $\exp\left(i\vec{b}_{\vec{g}} \cdot \vec{r}\right)$, in which the periodic crystal potential was expanded, are characterized by the translation vectors $\vec{b}_{\vec{g}}$ of the reciprocal lattice. The relation between the reciprocal and direct lattices is such that the translation vectors of the reciprocal lattice define the plane waves that have the periodicity of the direct lattice. So we have

$$\exp\left[i\vec{b}_{\vec{g}} \cdot \left(\vec{r} + \vec{a}_{\vec{n}}\right)\right] = \exp\left(i\vec{b}_{\vec{g}} \cdot \vec{r}\right), \qquad (5.31)$$

which, at the same time, guarantees the periodicity of the crystal potential $V(\vec{r})$.

It can be easily proved that

$$\vec{a}_i \cdot \vec{b}_k = 2\pi \delta_{ik} = \begin{cases} 0, & \text{for } i \neq k \\ 2\pi, & \text{for } i = k \end{cases} \qquad (5.32)$$

and we will use this property of vectors $\vec{a}_i$ and $\vec{b}_k$ to calculate the volume of the unit cell of the reciprocal lattice. Since $\vec{b}_3 = 2\pi (\vec{a}_1 \times \vec{a}_2)/\Omega_0$, we have

$$V = (\vec{b}_1 \times \vec{b}_2) \cdot \vec{b}_3 = \frac{2\pi}{\Omega_0}(\vec{b}_1 \times \vec{b}_2) \cdot (\vec{a}_1 \times \vec{a}_2)$$

$$= \frac{2\pi}{\Omega_0} \left[\left(\vec{b}_1 \cdot \vec{a}_1\right)\left(\vec{b}_2 \cdot \vec{a}_2\right) - \left(\vec{b}_1 \cdot \vec{a}_2\right)\left(\vec{b}_2 \cdot \vec{a}_1\right) \right] = \frac{(2\pi)^3}{\Omega_0},$$

$$(5.33)$$

where we used the identity

$$\left(\vec{A} \times \vec{B}\right) \cdot \left(\vec{C} \times \vec{D}\right) = \left(\vec{A} \cdot \vec{C}\right)\left(\vec{B} \cdot \vec{D}\right) - \left(\vec{A} \cdot \vec{D}\right)\left(\vec{B} \cdot \vec{C}\right), \quad (5.34)$$

which is true for any three vectors, and in the last step, we used the relations given by Eq. 5.32. We can see from Eq. 5.33 that the volume of the unit cell of the reciprocal lattice is equal to the inverse of the volume of the unit cell of the direct lattice multiplied by factor $(2\pi)^3$.

5.3 Examples of Reciprocal Lattices

5.3.1 *Introduction*

We will now give some examples of reciprocal lattices. As a first example, we will consider the triclinic lattice.

5.3.2 *Reciprocal of the Triclinic Lattice*

We can see in Fig. 5.4 that the reciprocal of a triclinic lattice is also triclinic. Each of the $\vec{b}_i$ vectors (which define a primitive unit cell of the reciprocal lattice) is orthogonal to the plane defined by two of the three vectors $\vec{a}_1, \vec{a}_2, \vec{a}_3$, which are the generators of the direct lattice. It should be noted that the dimensions of the cells in Fig. 5.4 are not comparable since the unit of a_i is meter and that of b_i is inverse meter.

We will consider now two more lattices which belong to the cubic crystal system.

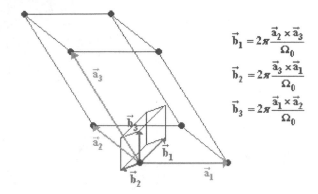

$$\vec{b}_1 = 2\pi \frac{\vec{a}_2 \times \vec{a}_3}{\Omega_0}$$

$$\vec{b}_2 = 2\pi \frac{\vec{a}_3 \times \vec{a}_1}{\Omega_0}$$

$$\vec{b}_3 = 2\pi \frac{\vec{a}_1 \times \vec{a}_2}{\Omega_0}$$

Figure 5.4 Primitive unit cell of the reciprocal of the triclinic lattice. In the figure, we show also the primitive unit cell for the direct lattice.

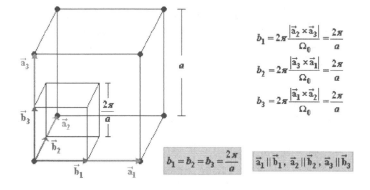

$$b_1 = 2\pi \frac{|\vec{a}_2 \times \vec{a}_3|}{\Omega_0} = \frac{2\pi}{a}$$

$$b_2 = 2\pi \frac{|\vec{a}_3 \times \vec{a}_1|}{\Omega_0} = \frac{2\pi}{a}$$

$$b_3 = 2\pi \frac{|\vec{a}_1 \times \vec{a}_2|}{\Omega_0} = \frac{2\pi}{a}$$

$$b_1 = b_2 = b_3 = \frac{2\pi}{a} \qquad \vec{a}_1 \| \vec{b}_1, \ \vec{a}_2 \| \vec{b}_2, \ \vec{a}_3 \| \vec{b}_3$$

Figure 5.5 Cubic *P* unit cells of the *sc* lattice and its reciprocal lattice.

5.3.3 *Reciprocal of the Simple Cubic Lattice*

The primitive unit cell of the reciprocal of the *sc* lattice has a cubic shape, so this reciprocal lattice is also simple cubic. This is shown in Fig. 5.5. The volume of the unit cell of the reciprocal lattice is given by the expression

$$V = b_1 \cdot b_2 \cdot b_3 = \frac{(2\pi)^3}{a^3} = \frac{(2\pi)^3}{\Omega_0}, \tag{5.35}$$

where Ω_0 is the volume of the primitive unit cell for the direct lattice.

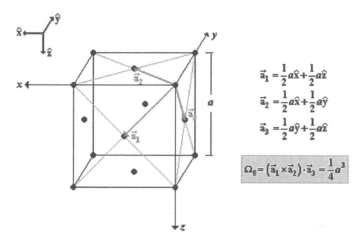

$$\vec{a}_1 = \frac{1}{2}a\hat{x} + \frac{1}{2}a\hat{z}$$

$$\vec{a}_2 = \frac{1}{2}a\hat{x} + \frac{1}{2}a\hat{y}$$

$$\vec{a}_3 = \frac{1}{2}a\hat{y} + \frac{1}{2}a\hat{z}$$

$$\Omega_0 = (\vec{a}_1 \times \vec{a}_2) \cdot \vec{a}_3 = \frac{1}{4}a^3$$

Figure 5.6 Cubic F unit cell of the *fcc* lattice and the primitive translation vectors $\vec{a}_1, \vec{a}_2, \vec{a}_3$ that define a rhombohedral P unit cell for this lattice.

5.3.4 *Reciprocal of the Face Centered Cubic Lattice*

Let us first remind some basics about the *fcc* lattice. The primitive translation vectors that define a rhombohedral P unit cell for the *fcc* lattice are shown in Fig. 5.6. The volume of the rhombohedral P unit cell is also given in this figure. The three primitive translation vectors for the reciprocal of the *fcc* lattice are calculated below. We have

$$\vec{b}_1 = 2\pi \frac{\vec{a}_2 \times \vec{a}_3}{\Omega_0} = \frac{2\pi}{\frac{1}{4}a^3} \begin{vmatrix} \hat{x} & \hat{y} & \hat{z} \\ \frac{1}{2}a & \frac{1}{2}a & 0 \\ 0 & \frac{1}{2}a & \frac{1}{2}a \end{vmatrix} = \frac{2\pi}{a}\hat{x} - \frac{2\pi}{a}\hat{y} + \frac{2\pi}{a}\hat{z}$$

$$\vec{b}_2 = 2\pi \frac{\vec{a}_3 \times \vec{a}_1}{\Omega_0} = \frac{2\pi}{\frac{1}{4}a^3} \begin{vmatrix} \hat{x} & \hat{y} & \hat{z} \\ 0 & \frac{1}{2}a & \frac{1}{2}a \\ \frac{1}{2}a & 0 & \frac{1}{2}a \end{vmatrix} = \frac{2\pi}{a}\hat{x} + \frac{2\pi}{a}\hat{y} - \frac{2\pi}{a}\hat{z}$$

$$\vec{b}_3 = 2\pi \frac{\vec{a}_1 \times \vec{a}_2}{\Omega_0} = \frac{2\pi}{\frac{1}{4}a^3} \begin{vmatrix} \hat{x} & \hat{y} & \hat{z} \\ \frac{1}{2}a & 0 & \frac{1}{2}a \\ \frac{1}{2}a & \frac{1}{2}a & 0 \end{vmatrix} = -\frac{2\pi}{a}\hat{x} + \frac{2\pi}{a}\hat{y} + \frac{2\pi}{a}\hat{z}$$

$$\tag{5.36}$$

Using the above expressions for $\vec{b}_i$ it is easy to show that

$$b_1 = b_2 = b_3 = \sqrt{3 \left(\frac{2\pi}{a}\right)^2} = \frac{2\pi\sqrt{3}}{a} \qquad (5.37)$$

and also

$$\sphericalangle\left(\vec{b}_1, \vec{b}_2\right) = \sphericalangle\left(\vec{b}_2, \vec{b}_3\right) = \sphericalangle\left(\vec{b}_3, \vec{b}_1\right). \qquad (5.38)$$

This means that a primitive unit cell of the reciprocal lattice has also a rhombohedral shape. If we now compare the expressions for vectors $\vec{b}_1$, $\vec{b}_2$, $\vec{b}_3$, given by Eqs. 5.36, with the expressions for the vectors $\vec{a}_1$, $\vec{a}_2$, $\vec{a}_3$ given in Section 2.9 for the *bcc* lattice (see the equations in Fig. 2.26), then we find that *bcc* is the reciprocal lattice of the *fcc* lattice with lattice constant $4\pi/a$ (see Fig. 5.7). It is also true that *fcc* is the reciprocal lattice of the *bcc* lattice (see Exercise 5.3). The correspondence between such types of centering in direct and reciprocal lattices is quite general, which means that the same occurs in the case of *oI*, *oF*, and *tI* direct lattices and their reciprocal lattices. The reciprocal to the direct *tI* lattice is the *tF* lattice, which is equivalent to the reciprocal *tI* lattice. The correspondence between different types of centering in direct and

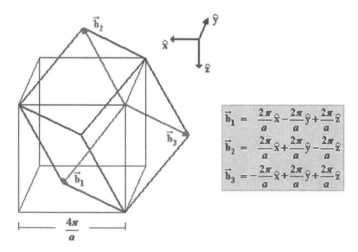

Figure 5.7 Rhombohedral P unit cell for the *bcc* reciprocal lattice. The cell is defined by primitive translation vectors $\vec{b}_1$, $\vec{b}_2$, $\vec{b}_3$ with components given in the figure.

Table 5.1 Correspondence between types of centering in direct and their reciprocal lattices

Type of direct lattice	Type of reciprocal lattice
Primitive	Primitive
A-centered	A-centered
B-centered	B-centered
C-centered	C-centered
Body centered	F-centered
F-centered	Body centered

reciprocal lattices is given in Table 5.1. The two lattices, direct and its reciprocal, have the same point-group symmetry.

In Fig. 5.7 we have placed the rhombohedral P unit cell and the cubic I unit cell for the reciprocal of the fcc lattice. Note that vectors $\vec{b}_i$ go from a vertex of the cube shown in the figure towards three centers of the cubes adjacent to this cube (the ones that share a common vertex with this cube).

Problems

Exercise 5.1 Equation 5.22 is reduced to the following form in the case of a two-dimensional lattice,

$$\vec{b} \cdot \vec{a}_n = n_1 \left(\vec{b} \cdot \vec{a}_1\right) + n_2 \left(\vec{b} \cdot \vec{a}_2\right) = (\text{integer number}) \cdot 2\pi, \quad (5.39)$$

for all possible n_1 and $n_2 \in \mathbb{Z}$. The above equation represents the condition for translation vectors $\vec{b}$ of the reciprocal of a two-dimensional crystal lattice generated by primitive vectors $\vec{a}_1$ and $\vec{a}_2$. The condition given by Eq. 5.39 is satisfied only if

$$\begin{cases} \vec{b} \cdot \vec{a}_1 = 2\pi \cdot g_1 \\ \vec{b} \cdot \vec{a}_2 = 2\pi \cdot g_2 \end{cases}, \quad \text{where } g_1 \text{ and } g_2 \in \mathbb{Z}. \quad (5.40)$$

From Eqs. 5.40, we deduce that there are vectors $\vec{b}$ that are orthogonal to $\vec{a}_1$ or $\vec{a}_2$. Two such vectors are shown in Fig. 5.8. Let us denote those two vectors, which are primitive translation vectors, as $\vec{b}_1$ and $\vec{b}_2$.

(a) Show that the translation vector $\vec{b}$ of the reciprocal lattice has the following expression

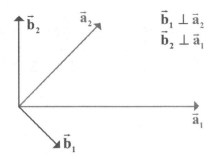

Figure 5.8 Vectors $\vec{a}_1$, $\vec{a}_2$ and $\vec{b}_1$, $\vec{b}_2$ that generate the two-dimensional direct and reciprocal lattices, respectively.

$$\vec{b} = g_1\vec{b}_1 + g_2\vec{b}_2, \qquad (5.41)$$

where g_1 and g_2 are specified by Eqs. 5.40, and $\vec{b}_1$ and $\vec{b}_2$ are orthogonal to $\vec{a}_2$ and $\vec{a}_1$, respectively. To do so, express the translation vector $\vec{b}$ as a linear combination of unit vectors $\hat{b}_1$ and $\hat{b}_2$ with coefficients α and β

$$\vec{b} = \alpha\hat{b}_1 + \beta\hat{b}_2, \qquad (5.42)$$

then find those coefficients and the expressions for the primitive translation vectors $\vec{b}_1$ and $\vec{b}_2$.

(b) For each of the five lattices existing in two dimensions, draw the primitive unit cell for its reciprocal lattice (defined by vectors $\vec{b}_1$ and $\vec{b}_2$ specified in (a)) together with the primitive unit cell for the direct lattice shown in Fig. 1.19 of Section 1.3.

Exercise 5.2 Show that the primitive translation vectors $\vec{a}_1$, $\vec{a}_2$ and $\vec{b}_1$, $\vec{b}_2$ for a two-dimensional crystal lattice and its reciprocal, respectively, satisfy Eq. 5.32. To do so, solve first point (a) in Exercise 5.1 in order to have the expressions for $\vec{b}_1$ and $\vec{b}_2$.

Exercise 5.3

(a) Show that the reciprocal of the *bcc* lattice (with lattice constant a) is the *fcc* lattice (with lattice constant $4\pi/a$).

(b) Using the primitive translation vectors for the *fcc* reciprocal lattice, obtained in (a), draw the primitive unit cell inside the cubic F unit cell for this lattice.

Exercise 5.4 Prove that the reciprocal of a reciprocal lattice is its direct lattice. To do so, substitute the expressions for $\vec{b}_1, \vec{b}_2, \vec{b}_3$ given by Eqs. 5.29 into the expressions for the primitive translation vectors of the reciprocal of a reciprocal lattice given by

$$2\pi \frac{(\vec{b}_2 \times \vec{b}_3)}{V}, \quad 2\pi \frac{(\vec{b}_3 \times \vec{b}_1)}{V}, \quad 2\pi \frac{(\vec{b}_1 \times \vec{b}_2)}{V},$$

where V is the volume of the primitive unit cell of the reciprocal lattice. To simplify the result, make use of the vector identity

$$\vec{A} \times \left(\vec{B} \times \vec{C} \right) = \vec{B} \left(\vec{A} \cdot \vec{C} \right) - \vec{C} \left(\vec{A} \cdot \vec{B} \right) \tag{5.43}$$

and Eq. 5.33.

Chapter 6

Direct and Reciprocal Lattices

6.1 Introduction

A three-dimensional Bravais lattice may be seen as a set of two-dimensional lattices, whose planes are parallel to each other and equally spaced. Each of these planes represents a lattice plane of the three-dimensional Bravais lattice. The way of seeing a three-dimensional lattice as a set of two-dimensional lattices is not unique. A set of parallel, equally spaced lattice planes is known as a family of lattice planes. The orientation of the planes belonging to each family is given by the so-called *Miller indices*. We will show in this chapter that the Miller indices represent the components of a translation vector of the reciprocal lattice which is orthogonal to the family of the lattice planes labeled with these indices. In the next section, we will learn how to obtain the Miller indices.

6.2 Miller Indices

We will show, first, examples of lattice planes in a given Bravais lattice. A lattice plane is defined by at least three non-collinear

Basic Elements of Crystallography (2nd Edition)
Nevill Gonzalez Szwacki and Teresa Szwacka
Copyright © 2016 Pan Stanford Publishing Pte. Ltd.
ISBN 978-981-4613-57-6 (Hardcover), 978-981-4613-58-3 (eBook)
www.panstanford.com

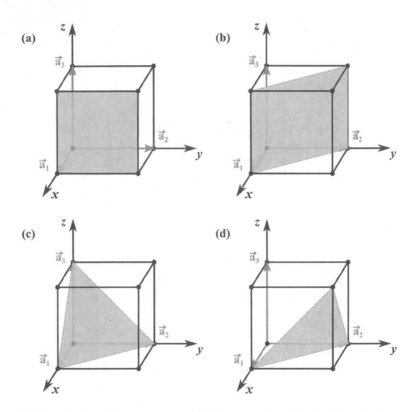

Figure 6.1 Four lattice planes with different orientations in the *sc* lattice.

lattice points. In Fig. 6.1 we can see four lattice planes with different orientations in the *sc* lattice.

Let us now introduce the Miller indices. They specify the orientation of a Bravais lattice plane (or the family of planes) in a very useful manner, what we will see later. The Miller indices, h, k, and l, can be obtained as follows:

(a) From the family of lattice planes that are parallel to each other, we select a plane that crosses the lattice axes (defined by the primitive translation vectors $\vec{a}_1$, $\vec{a}_2$, and $\vec{a}_3$) in the lattice points. The position vectors of these points, given in general in a non-orthogonal reference system with axes along $\vec{a}_1$, $\vec{a}_2$, $\vec{a}_3$ (see Fig. 6.2), are $\vec{r}_1 = s_1\vec{a}_1$, $\vec{r}_2 = s_2\vec{a}_2$, $\vec{r}_3 = s_3\vec{a}_3$, where s_1, s_2, and $s_3 \in \mathbb{Z}$.

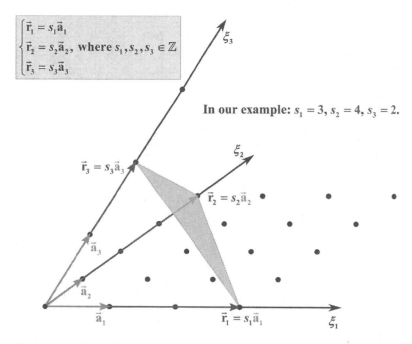

$$\begin{cases} \vec{r}_1 = s_1\vec{a}_1 \\ \vec{r}_2 = s_2\vec{a}_2, \text{ where } s_1, s_2, s_3 \in \mathbb{Z} \\ \vec{r}_3 = s_3\vec{a}_3 \end{cases}$$

In our example: $s_1 = 3$, $s_2 = 4$, $s_3 = 2$.

Figure 6.2 Three-dimensional crystal lattice generated by the primitive translation vectors $\vec{a}_1, \vec{a}_2, \vec{a}_3$. The lattice plane shown in the figure intersects the axes ξ_1, ξ_2, and ξ_3 in the lattice points.

(b) Next, we take the inverse values of the numbers s_1, s_2, and s_3 and reduce them to the smallest integers with the same ratio, namely,

$$\frac{1}{s_1} : \frac{1}{s_2} : \frac{1}{s_3} = h : k : l. \tag{6.1}$$

Obtained in this way, the integer numbers h, k, and l with no common factors are known as the Miller indices, which placed in parenthesis, (hkl), denote a single lattice plane or a family of lattice planes parallel to each other.

As a first example, we will use Eq. 6.1 for the case shown in Fig. 6.2. For this case

$$\frac{1}{s_1} : \frac{1}{s_2} : \frac{1}{s_3} = \frac{1}{3} : \frac{1}{4} : \frac{1}{2} = 4 : 3 : 6 = h : k : l, \tag{6.2}$$

that is to say, the Miller indices of the plane represented in Fig. 6.2 are 4, 3, and 6 and the plane is specified by (436). Let us now describe the cases shown in Figs. 6.1a–d. In these figures, we find four lattice planes in the *sc* lattice. The plane shaded in Fig. 6.1a includes a cube face. It intersects only the x axis in the point $\vec{r}_1 = 1\vec{a}_1$ (the other intercepts are at the "infinity"); thus, Eq. 6.1 turns to the following form for this case:

$$\frac{1}{1} : \frac{1}{\infty} : \frac{1}{\infty} = 1 : 0 : 0 \Rightarrow (hkl) = (100). \tag{6.3}$$

In the similar way, we obtain the Miller indices for the rest of the planes shown in Fig. 6.1, namely,

$$\frac{1}{1} : \frac{1}{1} : \frac{1}{\infty} = 1 : 1 : 0 \Rightarrow (hkl) = (110),$$

$$\frac{1}{1} : \frac{1}{1} : \frac{1}{1} = 1 : 1 : 1 \Rightarrow (hkl) = (111), \text{ and}$$

$$\frac{1}{1} : \frac{1}{1} : \frac{1}{-1} = 1 : 1 : \bar{1} \Rightarrow (hkl) = \left(11\bar{1}\right) \tag{6.4}$$

for Figs. 6.1b–d, respectively. The shaded plane in Fig. 6.1d intersects the z axis in the point $\vec{r}_3 = -1\vec{a}_3$ (see also Fig. 6.3). Due to the

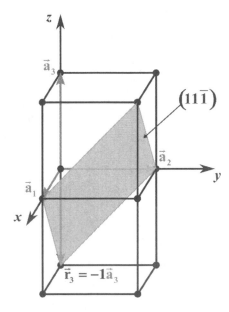

Figure 6.3 A complementary figure to Fig. 6.1d. In this figure is indicated the lattice point where the $\left(11\bar{1}\right)$ plane crosses the z-axis defined by the translation vector $\vec{a}_3$.

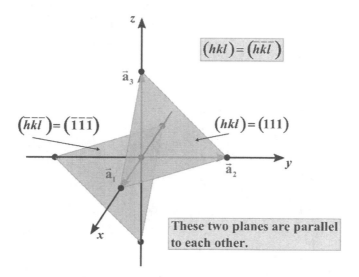

Figure 6.4 The (111) and $(\bar{1}\bar{1}\bar{1})$ planes in the *sc* lattice.

convention, the negative Miller indices are written with a bar, which means, instead of -1 we have $\bar{1}$. From the definition of the Miller indices, it is easy to see that

$$\left(\bar{h}\bar{k}\bar{l}\right) = (hkl). \qquad (6.5)$$

The equality 6.5 is illustrated in Fig. 6.4 for the case of the (111) and $(\bar{1}\bar{1}\bar{1})$ planes in the *sc* lattice that are parallel to each other.

As we have learned, the Miller indices are used to identify a single lattice plane and also a family of planes parallel to each other. For a set of lattice planes (or a set of families of parallel lattice planes) that are equivalent by symmetry of the lattice, there is also a notation. Let us illustrate this on the example of the planes which include the three faces of the cubic unit cell for the *sc* lattice shown in Fig. 6.5. "Curly" brackets, $\{100\}$, designate the (100) plane together with the (010) and (001) planes that are equivalent by lattice symmetry to it. In general, the notation $\{hkl\}$ refers to the (hkl) planes and all other families of lattice planes that are equivalent to them by symmetry of the lattice.

Now, we will introduce a convention to specify a direction in a three-dimensional direct lattice. This was considered already for the two-dimensional case in Section 1.4. Such direction can be identified

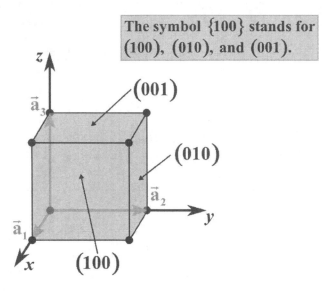

Figure 6.5 Three lattice planes in the *sc* lattice that are equivalent by symmetry of the lattice.

by the three components of vector $\vec{a}_{\bar{1}}$, which is the shortest one in this direction (see Fig. 6.6). In order to determine the components of this vector, we can take a vector $\vec{R}$ defined by two lattice points in the direction in consideration and make the reductions to the three smallest integers. For example, in Fig. 6.6 we have proposed $\vec{R} = 3\vec{a}_1 + 0\vec{a}_2 + 3\vec{a}_3$; next, we take the integer numbers that multiply the primitive translation vectors $\vec{a}_1$, $\vec{a}_2$, $\vec{a}_3$ and then reduce them to the smallest integers having the same ratio: $3 : 0 : 3 = 1 : 0 : 1$. In this manner, we can obtain the components of the vector $\vec{a}_{\bar{1}} = 1\vec{a}_1 + 0\vec{a}_2 + 1\vec{a}_3$, which is the shortest one in the lattice direction in consideration. The notation [101], with square brackets instead of round brackets, is used to specify the lattice direction shown in Fig. 6.6. In general, the notation $[l_1 l_2 l_3]$ denotes a crystal lattice direction with the shortest translation vector $\vec{a}_{\bar{1}} = l_1 \vec{a}_1 + l_2 \vec{a}_2 + l_3 \vec{a}_3$.

All directions that are equivalent to $[l_1 l_2 l_3]$ by lattice symmetry are denoted with the symbol $\langle l_1 l_2 l_3 \rangle$. Figure 6.7 shows an example of three equivalent directions, [100], [010], and [001], in the *sc* lattice. The set of these directions, together with [$\bar{1}$00], [0$\bar{1}$0], and [00$\bar{1}$], is denoted by $\langle 100 \rangle$.

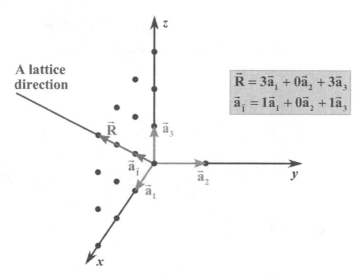

$$\vec{R} = 3\vec{a}_1 + 0\vec{a}_2 + 3\vec{a}_3$$
$$\vec{a}_{\bar{1}} = 1\vec{a}_1 + 0\vec{a}_2 + 1\vec{a}_3$$

Figure 6.6 A lattice direction. The vector $\vec{a}_{\bar{1}}$ is the shortest one in this direction.

6.3 Application of Miller Indices

We are going to describe now some of the properties of Bravais lattices with the aid of Miller indices.

Property 1

First, we will look for the positions of three points that define a lattice plane which is the closest to the plane that passes through the origin of the non-orthogonal reference system, defined by the primitive translation vectors $\vec{a}_1, \vec{a}_2, \vec{a}_3$ in Fig. 6.8. The vectors $\vec{r}_1, \vec{r}_2,$ $\vec{r}_3$ shown in Fig. 6.8 give the positions of three lattice points. The lattice plane that intersects the $\xi_1, \xi_2,$ and ξ_3 axes in these points has the following Miller indices

$$h : k : l = \frac{1}{3} : \frac{1}{1} : \frac{1}{1} = 1 : 3 : 3. \tag{6.6}$$

The other (133) lattice plane shown in Fig. 6.8 is the closest plane to the one that passes through the origin. This plane crosses the lattice

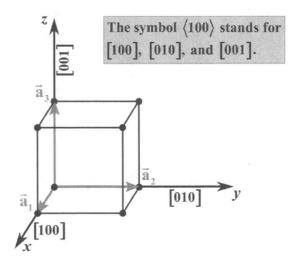

The symbol ⟨100⟩ stands for [100], [010], and [001].

Figure 6.7 Three directions equivalent by lattice symmetry in the *sc* lattice.

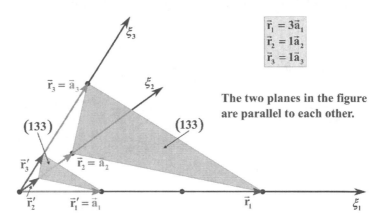

$\bar{r}_1 = 3\bar{a}_1$
$\bar{r}_2 = 1\bar{a}_2$
$\bar{r}_3 = 1\bar{a}_3$

The two planes in the figure are parallel to each other.

Figure 6.8 Two (133) lattice planes in a lattice generated by the primitive translation vectors $\vec{a}_1$, $\vec{a}_2$, $\vec{a}_3$.

axes in the points given by vectors

$$
\begin{cases}
\vec{r}_1' = 1\vec{a}_1 = \dfrac{\vec{a}_1}{h} \\[2mm]
\vec{r}_2' = \dfrac{1}{3}\vec{a}_2 = \dfrac{\vec{a}_2}{k} \\[2mm]
\vec{r}_3' = \dfrac{1}{3}\vec{a}_3 = \dfrac{\vec{a}_3}{l}
\end{cases}
. \tag{6.7}
$$

Therefore, the plane intersects the ξ_1, ξ_2, and ξ_3 axes at the points $\vec{a}_1/h$, $\vec{a}_2/k$, and $\vec{a}_3/l$, respectively. This general statement can be deduced from the geometric considerations related to the intercepts with the axes of equidistant and parallel to each other (hkl) planes.

Property 2

Next, we will show that the reciprocal lattice vector $\vec{b}_{hkl} = h\vec{b}_1 + k\vec{b}_2 + l\vec{b}_3$ is perpendicular to the (hkl) direct lattice plane. To demonstrate that, from all the planes of the (hkl) family, we will take the one that crosses the lattice axes (defined by the primitive translation vectors $\vec{a}_1$, $\vec{a}_2$, $\vec{a}_3$) in the points given by $\vec{a}_1/h$, $\vec{a}_2/k$, $\vec{a}_3/l$. The non-collinear vectors $(\vec{a}_2/k - \vec{a}_1/h)$ and $(\vec{a}_2/k - \vec{a}_3/l)$ are on a (hkl) plane that is the closest to the origin, what is illustrated for the case of the (623) plane in Fig. 6.9. It will suffice to show that $\vec{b}_{hkl} \perp (\vec{a}_2/k - \vec{a}_1/h)$ and $\vec{b}_{hkl} \perp (\vec{a}_2/k - \vec{a}_3/l)$ to be able to say that $\vec{b}_{hkl}$ is orthogonal to the family of (hkl) planes. We will calculate the following scalar products for this purpose

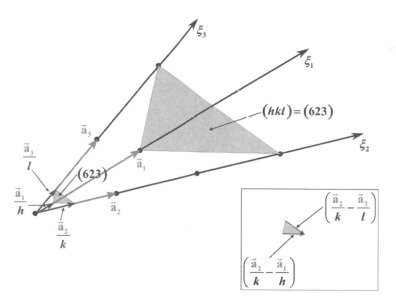

Figure 6.9 Two (623) lattice planes in a lattice generated by the primitive translation vectors $\vec{a}_1$, $\vec{a}_2$, $\vec{a}_3$.

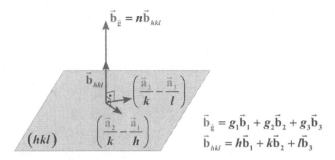

Figure 6.10 The direction given by the vector $\vec{b}_{hkl}$ in the reciprocal lattice is orthogonal to the family of (hkl) direct lattice planes.

$$\vec{b}_{hkl} \cdot \left(\frac{\vec{a}_2}{k} - \frac{\vec{a}_1}{h}\right) = \left(h\vec{b}_1 + k\vec{b}_2 + l\vec{b}_3\right) \cdot \left(\frac{\vec{a}_2}{k} - \frac{\vec{a}_1}{h}\right) = 2\pi - 2\pi = 0$$

$$\vec{b}_{hkl} \cdot \left(\frac{\vec{a}_2}{k} - \frac{\vec{a}_3}{l}\right) = \left(h\vec{b}_1 + k\vec{b}_2 + l\vec{b}_3\right) \cdot \left(\frac{\vec{a}_2}{k} - \frac{\vec{a}_3}{l}\right) = 2\pi - 2\pi = 0.$$

$$(6.8)$$

In this manner, we have demonstrated that the vector $\vec{b}_{hkl}$ is indeed orthogonal to the family of (hkl) lattice planes. Concluding, we can say that the Miller indices h, k, and l are the coordinates of the shortest reciprocal lattice vector $\vec{b}_{hkl} = h\vec{b}_1 + k\vec{b}_2 + l\vec{b}_3$, which is orthogonal to the (hkl) planes in the direct lattice. Of course, any vector that is a multiple of the $\vec{b}_{hkl}$ vector, $n\vec{b}_{hkl}$, is also orthogonal to the (hkl) plane. Certainly, if g_1, g_2, and g_3, specified in Fig. 6.10, fulfill the relation $g_1 : g_2 : g_3 = h : k : l$, that is to say

$$\frac{g_1}{h} = \frac{g_2}{k} = \frac{g_3}{l} = n \in \mathbb{Z},$$

$$(6.9)$$

then $\vec{b}_{\vec{g}} = n\vec{b}_{hkl}$ and the $\vec{b}_{\vec{g}}$ vector is orthogonal to the (hkl) plane. To conclude, we can say that the direction defined by the vector $\vec{b}_{hkl}$ in the reciprocal lattice corresponds to the (hkl) planes in the direct lattice. The vector $\vec{b}_{hkl}$ can be used to define a unit vector that is orthogonal to (hkl) planes

$$\hat{n}_{hkl} = \frac{\vec{b}_{hkl}}{b_{hkl}}.$$

$$(6.10)$$

The unit vector given by Eq. 6.10 specifies the orientation of a lattice plane denoted (hkl).

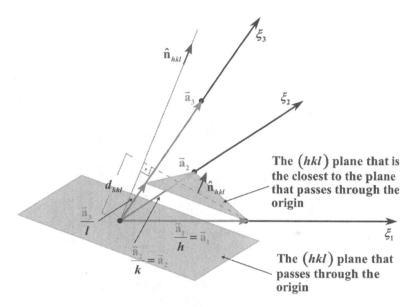

Figure 6.11 Two consecutive (hkl) planes. The direction orthogonal to these planes, defined by unit vector $\hat{n}_{hkl}$, is shown. The distance, d_{hkl}, between these planes is also indicated.

Property 3

At last, we will calculate the distance between two consecutive (hkl) planes, d_{hkl}, using the vector $\hat{n}_{hkl}$. To obtain the d_{hkl} parameter it is sufficient to project, for example, $\vec{a}_1/h$ in the direction orthogonal to the (hkl) planes (as it is done in Fig. 6.11), that is to say

$$d_{hkl} = \frac{\vec{a}_1}{h} \cdot \hat{n}_{hkl} = \frac{\vec{a}_1}{h} \cdot \frac{\left(h\vec{b}_1 + k\vec{b}_2 + l\vec{b}_3 \right)}{b_{hkl}} = \frac{2\pi}{b_{hkl}}. \qquad (6.11)$$

This means that two consecutive planes of the family of (hkl) planes are at a distance which is equal to the inverse of the modulus of the $\vec{b}_{hkl}$ vector multiplied by 2π.

As an example, let us now apply the formula that we obtained for d_{hkl} for the *sc* lattice. We will start with the family of the (100) lattice planes. The information about the vectors that generate the direct, $\vec{a}_1, \vec{a}_2, \vec{a}_3$, and the reciprocal, $\vec{b}_1, \vec{b}_2, \vec{b}_3$, lattices is given in Fig. 6.12. We can see in this figure that $\vec{b}_1 \parallel \vec{a}_1$, $\vec{b}_2 \parallel \vec{a}_2$, and $\vec{b}_3 \parallel \vec{a}_3$. So the vector $\vec{b}_{100} = 1\vec{b}_1 + 0\vec{b}_2 + 0\vec{b}_3 = \vec{b}_1$ is indeed orthogonal to the (100) plane. The distance between two consecutive (100) planes, which include

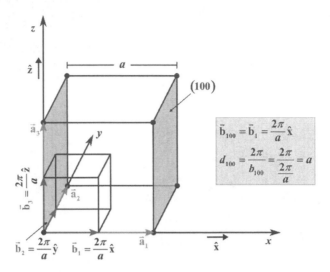

Figure 6.12 The (100) planes which include two cube faces that are parallel to each other in the direct *sc* lattice. The cubic *P* unit cell defined by primitive translation vectors $\vec{b}_1, \vec{b}_2, \vec{b}_3$ of the reciprocal lattice is also shown. The translation vector $\vec{b}_{100} = \vec{b}_1$ is orthogonal to the (100) planes.

two cube faces that are parallel to each other, is $d_{100} = 2\pi/b_{100} = a$, it means, it is equal to the cube edge lengths as it should be.

As a second example, we will consider the family of planes that are orthogonal to a body diagonal of the cube that represents the cubic *P* unit cell of the *sc* lattice. Two of these planes, denoted (111), are shown in Fig. 6.13. The figure shows also the cubic *P* unit cell of the reciprocal lattice, generated by the primitive translation vectors $\vec{b}_1, \vec{b}_2, \vec{b}_3$, which were obtained using the primitive translation vectors $\vec{a}_1, \vec{a}_2, \vec{a}_3$ and Eqs. 5.29. We can see in Fig. 6.13 that the vector $\vec{b}_{111}$ is parallel to the body diagonals of both cubes. So this vector is indeed orthogonal to the (111) direct lattice planes. The two (111) planes shown in Fig. 6.13 divide the body diagonal of the direct lattice unit cell in three segments of equal longitude. Thus the distance between these planes is equal to $1/3$ of the longitude of the diagonal of the cube

$$d_{111} = \frac{1}{3}\sqrt{3}a = \frac{\sqrt{3}}{3}a. \qquad (6.12)$$

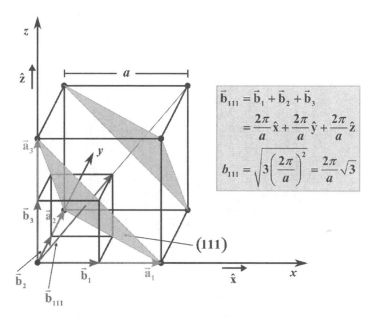

Figure 6.13 The cubic P unit cells of the sc direct lattice and the reciprocal to it. The vector $\vec{b}_{111}$ of the reciprocal lattice is orthogonal to the (111) planes in the direct lattice.

The distance between the (111) planes calculated using Eq. 6.11 is

$$d_{111} = \frac{2\pi}{b_{111}} = \frac{2\pi}{\frac{2\pi}{a}\sqrt{3}} = \frac{\sqrt{3}}{3}a. \qquad (6.13)$$

This result agrees with d_{111} given by Eq. 6.12, which we obtained from geometric considerations.

Problems

Exercise 6.1 In the sc lattice from Fig. 6.14

(a) draw five (111) lattice planes,
(b) draw all (221) lattice planes that contain at least two lattice points from the (010) front face of the large cube. Place additional points on the z axis if necessary.

Does the (221) lattice plane that is the closest to the origin belong to the set of planes specified in (b)?

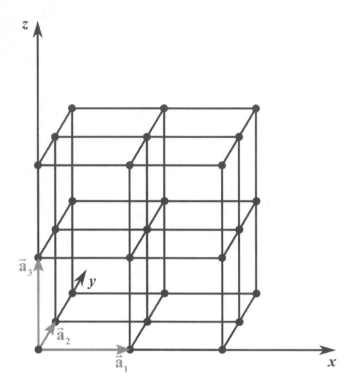

Figure 6.14   A simple cubic lattice, of lattice constant *a*, generated by the translation vectors $\vec{a}_1, \vec{a}_2, \vec{a}_3$.

Exercise 6.2   For the case (b) of Exercise 6.1 calculate the distance between two consecutive planes.

Exercise 6.3

(a) Without doing any calculations show that the consecutive lattice planes, orthogonal to body diagonals of cubic unit cells *P* and *F* of the *sc* and *fcc* lattices (of lattice constant *a*), respectively, are at the same distance from each other. Find this distance.

(b) Check your result, obtained for the case of the *fcc* lattice in (a), calculating the distance in consideration using Eq. 6.11. For that purpose use the primitive translation vectors that define the rhombohedral unit cell of the *fcc* lattice.

(c) Find the distance between the consecutive planes, orthogonal to a body diagonal of the cubic I unit cell of the *bcc* lattice with the same lattice constant a as has the *fcc* lattice in (b), and compare it with d_{111} calculated there.

Exercise 6.4 Figure 6.15 shows a two-dimensional lattice generated by the primitive translation vectors $\vec{a}_1$, $\vec{a}_2$ and four consecutive (41) planes in this lattice (Note, that in a two-dimensional lattice the planes are one-dimensional and are characterized by two Miller indices.). The Miller indices, 4 and 1, were calculated using the integer numbers s_1 and s_2 specified in Fig. 6.15

$$\frac{1}{s_1} : \frac{1}{s_2} = \frac{1}{1} : \frac{1}{4} = 4 : 1 = h : k. \tag{6.14}$$

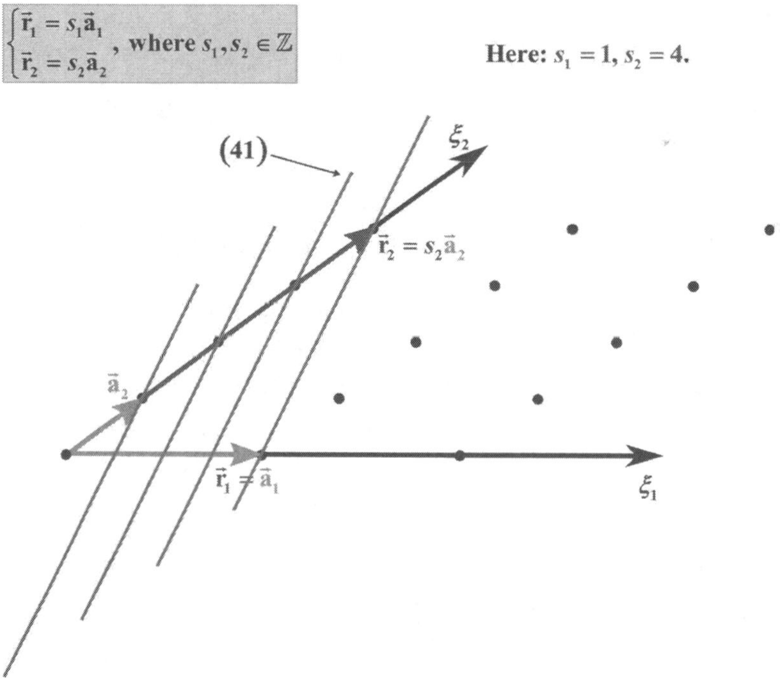

$$\begin{cases} \vec{r}_1 = s_1\vec{a}_1 \\ \vec{r}_2 = s_2\vec{a}_2 \end{cases}, \text{ where } s_1, s_2 \in \mathbb{Z}$$

Here: $s_1 = 1, s_2 = 4.$

Figure 6.15 Two-dimensional crystal lattice generated by the primitive translation vectors $\vec{a}_1$, $\vec{a}_2$. In the figure four consecutive (41) lattice planes are shown.

It is also true that

$$s_1 : s_2 = \frac{1}{h} : \frac{1}{k}, \tag{6.15}$$

thus the inverse of the Miller indices are at the same ratio as s_1 and s_2. Due to Eqs. 6.7 the (41) lattice plane, that is the closest to the plane that passes through the origin, intersects the axes ξ_1 and ξ_2 in the points $(1/h) \cdot \vec{a}_1$ and $(1/k) \cdot \vec{a}_2$, respectively. It means, the intercepts of this lattice plane with the axes ξ_1 and ξ_2 are $1/h$ and $1/k$, respectively. Figure 6.15 shows also three (41) planes whose intercepts with the ξ_1 axis represent the multiples of the smallest intercept, $1/h$, and are not larger than the integer s_1, that is

$$2 \cdot \frac{1}{h}, \quad 3 \cdot \frac{1}{h}, \quad \text{and } 4 \cdot \frac{1}{h} = s_1. \tag{6.16}$$

In similar way are obtained the intercepts of these planes with the ξ_2 axis, which are

$$2 \cdot \frac{1}{k}, \quad 3 \cdot \frac{1}{k}, \quad \text{and } 4 \cdot \frac{1}{k} = s_2. \tag{6.17}$$

The plane with integer intercepts s_1 and s_2 is the one that we usually use to determine the Miller indices.

(a) Show at least one more lattice point in each of the (41) lattice planes from Fig. 6.15 that have only one lattice point in the figure.

(b) Find and draw (keeping proportions) the primitive translation vectors $\vec{b}_1$ and $\vec{b}_2$ of the reciprocal lattice and show graphically that the vector $\vec{b}_{41}$ is orthogonal to the (41) planes.

Chapter 7

X-Ray Diffraction

7.1 Introduction

The interaction of X-rays of the wavelength of the order of lattice constants and shorter ones, allow us to learn about the crystal structure of materials. In this chapter, we will introduce the principles of X-ray diffraction for the case of X-rays interacting with an ideal crystal structure; we will limit our considerations to the case when X-rays are scattered elastically. The scattering of X-rays by electrons from the atoms will be treated within the first-order Born approximation, and within the isolated-atom approximation (each atom contributes separately to the electron density).

7.2 Laue Equations

In what follows, we are going to consider X-ray (or neutron) scattering by atoms of a crystal whose atomic basis is composed of one atom. First, we will observe scattering by two atoms, which are at a distance equal to $|\vec{a}_{\vec{m}}|$, where $\vec{a}_{\vec{m}} = m_1\vec{a}_1 + m_2\vec{a}_2 + m_3\vec{a}_3$ ($\vec{a}_1$, $\vec{a}_2$, $\vec{a}_3$ are primitive lattice translation vectors and m_1, m_2, and

Basic Elements of Crystallography (2nd Edition)
Nevill Gonzalez Szwacki and Teresa Szwacka
Copyright © 2016 Pan Stanford Publishing Pte. Ltd.
ISBN 978-981-4613-57-6 (Hardcover), 978-981-4613-58-3 (eBook)
www.panstanford.com

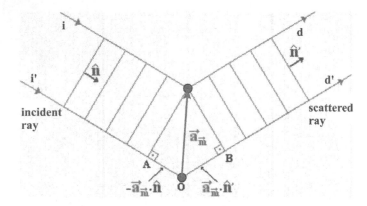

Figure 7.1 X-ray scattering by two atoms within a crystal structure with one-atom basis. One atom is located at the origin **O** and another one at a lattice point of position given by $\vec{a}_{\vec{m}}$. The unit vectors $\hat{n}$ and $\hat{n}'$ are parallel to the directions of the incident and scattered X-ray beams, respectively.

$m_3 \in \mathbb{Z}$). This is shown in Fig. 7.1. The X-rays disperse in all directions in each atom of the structure and a wave interference take place. In order for a constructive interference to occur the following has to be fulfilled

$$\text{AO} + \text{OB} = \text{integer number} \cdot \lambda, \tag{7.1}$$

where (AO + OB), defined in Fig. 7.1, represents the difference in lengths of the optical paths i–d and i'–d' of X-rays, while λ is the wavelength of the X-ray beam. From Eq. 7.1 and Fig. 7.1 we have

$$\text{AO} + \text{OB} = -\vec{a}_{\vec{m}} \cdot \hat{n} + \vec{a}_{\vec{m}} \cdot \hat{n}' = \vec{a}_{\vec{m}} \cdot (\hat{n}' - \hat{n})$$
$$= \text{integer number} \cdot \lambda, \tag{7.2}$$

which has to be fulfilled for each atom of the crystal, that is, for any $\vec{a}_{\vec{m}}$. Next, we have

$$\vec{a}_{\vec{m}} \cdot (\hat{n}' - \hat{n}) = (m_1\vec{a}_1 + m_2\vec{a}_2 + m_3\vec{a}_3) \cdot (\hat{n}' - \hat{n})$$
$$= m_1\vec{a}_1 \cdot (\hat{n}' - \hat{n}) + m_2\vec{a}_2 \cdot (\hat{n}' - \hat{n}) + m_3\vec{a}_3 \cdot (\hat{n}' - \hat{n})$$
$$= \text{integer number} \cdot \lambda. \tag{7.3}$$

Equation 7.3 can be rewritten as

$$m_1\frac{\vec{a}_1 \cdot (\hat{n}' - \hat{n})}{\lambda} + m_2\frac{\vec{a}_2 \cdot (\hat{n}' - \hat{n})}{\lambda} + m_3\frac{\vec{a}_3 \cdot (\hat{n}' - \hat{n})}{\lambda}$$
$$= \text{integer number}. \tag{7.4}$$

In order for this to happen for all possible m_1, m_2, and $m_3 \in \mathbb{Z}$, the following conditions have to be satisfied:

$$\begin{cases} \vec{a}_1 \cdot \dfrac{(\hat{n}' - \hat{n})}{\lambda} = g_1 \\[2mm] \vec{a}_2 \cdot \dfrac{(\hat{n}' - \hat{n})}{\lambda} = g_2 \\[2mm] \hat{a}_3 \cdot \dfrac{(\hat{n}' - \hat{n})}{\lambda} = g_3 \end{cases} \tag{7.5}$$

where g_1, g_2, and $g_3 \in \mathbb{Z}$. Reorganizing the three equations and multiplying them by 2π, we obtain

$$\begin{cases} \vec{a}_1 \cdot \left(\dfrac{2\pi\,\hat{n}'}{\lambda} - \dfrac{2\pi\,\hat{n}}{\lambda} \right) = 2\pi g_1 \\[2mm] \vec{a}_2 \cdot \left(\dfrac{2\pi\,\hat{n}'}{\lambda} - \dfrac{2\pi\,\hat{n}}{\lambda} \right) = 2\pi g_2 \\[2mm] \vec{a}_3 \cdot \left(\dfrac{2\pi\,\hat{n}'}{\lambda} - \dfrac{2\pi\,\hat{n}}{\lambda} \right) = 2\pi g_3 \end{cases} \tag{7.6}$$

From Eqs. 7.6, we see that vector $\left(\frac{2\pi\hat{n}'}{\lambda} - \frac{2\pi\hat{n}}{\lambda} \right)$ satisfies the same conditions as a vector of the reciprocal lattice $\vec{b}_{\vec{g}} = g_1 \vec{b}_1 + g_2 \vec{b}_2 + g_3 \vec{b}_3$ (g_1, g_2, and $g_3 \in \mathbb{Z}$), that is

$$\vec{a}_1 \cdot \vec{b}_{\vec{g}} = 2\pi g_1,\ \vec{a}_2 \cdot \vec{b}_{\vec{g}} = 2\pi g_2,\ \vec{a}_3 \cdot \vec{b}_{\vec{g}} = 2\pi g_3, \tag{7.7}$$

so

$$\frac{2\pi\,\hat{n}'}{\lambda} - \frac{2\pi\,\hat{n}}{\lambda} = \vec{b}_{\vec{g}}. \tag{7.8}$$

Recognizing that $2\pi\hat{n}/\lambda = \vec{k}$ and $2\pi\hat{n}'/\lambda = \vec{k}'$ ($\vec{k}$ and $\vec{k}'$ are the wave vectors of incident and scattered X-rays, respectively), we have finally

$$\vec{k}' - \vec{k} = \vec{b}_{\vec{g}}. \tag{7.9}$$

Equation 7.9 is illustrated in Fig. 7.2.

In the elastic dispersion, the difference in wave vectors $\vec{k}$ and $\vec{k}'$ of the incident and scattered X-ray beams has to be equal to a vector of the reciprocal lattice $\vec{b}_{\vec{g}}$, so that the constructive interference can occur. The conditions given in Eqs. 7.6 are called *Laue equations* and represent the conditions necessary for the constructive interference to occur. Of course, Eq. 7.9 is equivalent to Eqs. 7.6.

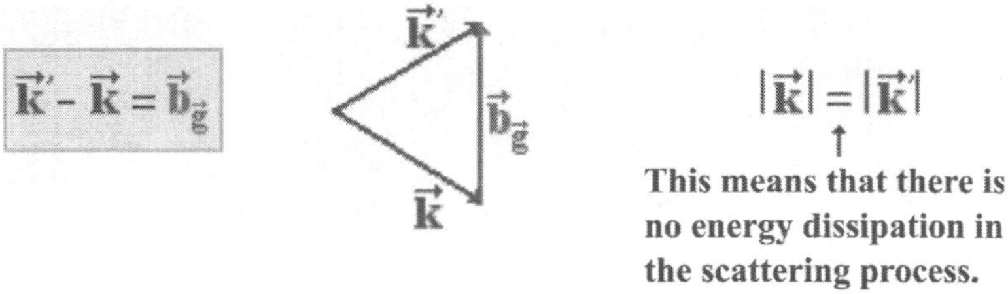

$$\vec{k}' - \vec{k} = \vec{b}_{\vec{g}}$$

$$|\vec{k}| = |\vec{k}'|$$

This means that there is no energy dissipation in the scattering process.

Figure 7.2 Geometric illustration of Eq. 7.9.

7.3 Ewald Construction and the Bragg Diffraction Formula

We will show now graphically the possible directions of the scattered X-ray beam in which the constructive interference can occur. Since $k' = k$, we have

$$\vec{k}'^2 = \left(\vec{k} + \vec{b}_g\right)^2 = \vec{k}^2 + \vec{b}_g^2 + 2\vec{k} \cdot \vec{b}_g = \vec{k}^2, \qquad (7.10)$$

which reduces to the formula

$$\vec{b}_g \cdot \left(\vec{k} + \frac{1}{2}\vec{b}_g\right) = 0. \qquad (7.11)$$

Equation 7.11 is illustrated graphically in Fig. 7.3. The condition given by Eq. 7.11 is equivalent to the conditions given by Eqs. 7.6 and 7.9. Next, we will place the drawing shown in Fig. 7.3 within the reciprocal lattice (to simplify, a two-dimensional case is considered), in the way that the end of vector $\vec{k}$ coincides with some node of the lattice. In Fig. 7.4, it is the node at **O**. From point I, shown in Fig. 7.4, we trace a circumference with a radius equal to $k = 2\pi/\lambda$. If the circumference passes through any other node of the

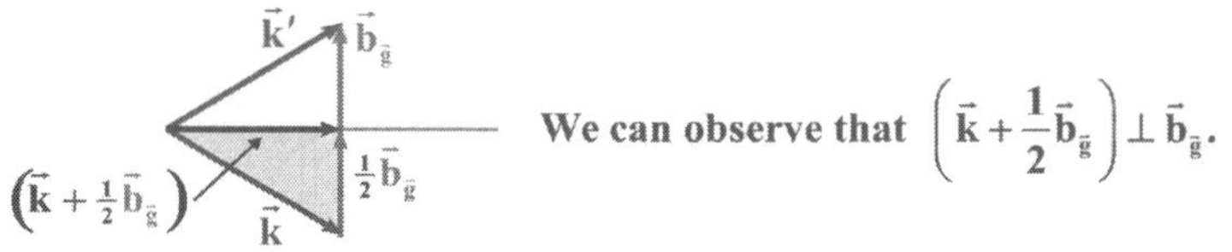

We can observe that $\left(\vec{k} + \frac{1}{2}\vec{b}_{\vec{g}}\right) \perp \vec{b}_{\vec{g}}$.

Figure 7.3 Graphical demonstration of the fact that vectors $\vec{b}_g$ and $\vec{k} + (1/2)\vec{b}_g$ are orthogonal to each other.

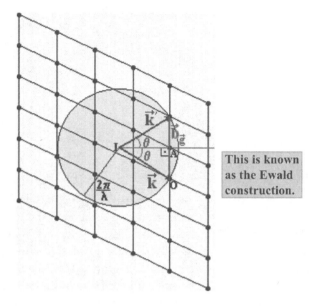

Figure 7.4 Ewald construction shown in a two-dimensional reciprocal lattice.

reciprocal lattice, the vector that goes from point I to this node is $\vec{k}'$ corresponding to the wave that propagates in the direction in which the constructive interference occurs. In this manner, we can visualize in the drawing the directions in which the constructive interference will occur. In a three-dimensional case, instead of the bisector of the segment defined by the vector $\vec{b}_{\vec{g}}$ (see Fig. 7.4), we have a plane orthogonal to this vector. This plane is one of the (hkl) planes of the direct lattice and vector $\vec{b}_{\vec{g}}$ is a multiple of the $\vec{b}_{hkl}$ vector

$$\vec{b}_{\vec{g}} = n\vec{b}_{hkl}. \tag{7.12}$$

Since the distance between neighboring planes, d_{hkl}, of the family of (hkl) planes is given by the expression

$$d_{hkl} = \frac{2\pi}{b_{hkl}} = \frac{2\pi n}{b_{\vec{g}}}, \tag{7.13}$$

$b_{\vec{g}}$ is expressed as a function of d_{hkl} as follows:

$$b_{\vec{g}} = \frac{2\pi n}{d_{hkl}}. \tag{7.14}$$

On the other hand, from the triangle AOI, shown in Fig. 7.4, we can see that $b_{\bar{g}}$ may be expressed as a function of θ:

$$\frac{b_{\bar{g}}}{2} = k \sin\theta = \frac{2\pi}{\lambda} \sin\theta. \tag{7.15}$$

Then from Eqs. 7.14 and 7.15 we have

$$\frac{2\pi n}{d_{hkl}} = \frac{4\pi}{\lambda} \sin\theta \tag{7.16}$$

and finally

$$2d_{hkl} \sin\theta = n\lambda. \tag{7.17}$$

Equation 7.17 represents the so-called *Bragg condition* (*Bragg's law*). It is easy to find the interpretation for the Bragg condition in the direct lattice. This is done in Fig. 7.5. In this figure, the scattering of X-rays by atoms within a crystal structure, in the direction in which there is a constructive interference, can be seen as a mirror reflection from a certain family of (hkl) lattice planes. The difference in lengths of the optical paths of the rays reflected by two consecutive planes, equal to $2d_{hkl} \sin\theta$, has to be a multiple of the X-ray wavelength (see Eq. 7.17). The Bragg condition is equivalent to the Laue condition and describes the constructive interference of X-rays with the parameters that are visualized within the direct lattice, while the Laue equations are linked directly to the reciprocal lattice.

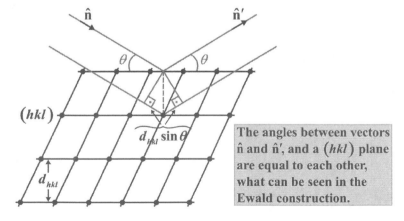

Figure 7.5 X-ray reflection from two consecutive (hkl) lattice planes.

In the deduction of the Laue or Bragg equations, the atoms are considered as point scatter centers located at the nodes of the lattice. However, the X-ray beam is scattered by the electron charge of atoms, which is distributed in a certain volume. To take this fact into account, we will introduce a parameter called the *atomic structure factor f*.

7.4 Atomic Structure Factor

As a first step, we will revise some elements of the quantum description of elastic collision phenomena, which includes X-ray scattering by a microscopic system. Suppose that an X-ray beam is scattered by a single scattering center (an electron, for example) that produces the potential $V(\vec{r})$ (see Fig. 7.6). The incoming ray is described by the plane wave $\phi_i(\vec{r}) = \exp\left(i\vec{k} \cdot \vec{r}\right)$. At a large distance from the scattering center the scattered X-rays propagate in all directions and are described by a spherical wave. The asymptotic expression of the wave function which describes the scattered wave in the $\hat{n}'$ direction reads, within the first-order Born approximation, as follows (the case of $\left|\vec{k}'\right| = \left|\vec{k}\right| = k$ is considered):

$$\psi_{\hat{n}'}(\vec{r}) = A(\hat{n}') \frac{\exp(ikr)}{r}, \tag{7.18}$$

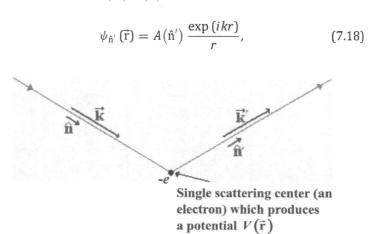

Single scattering center (an electron) which produces a potential $V(\vec{r})$

Figure 7.6 Scattering of X-rays by a single electron. The scattered ray propagates in any $\hat{n}'$ direction.

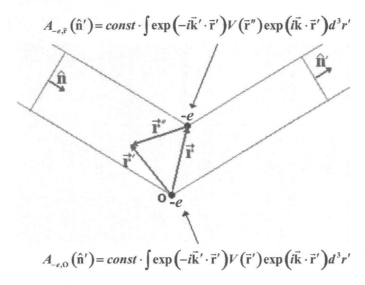

$$A_{-e,\vec{r}}(\hat{n}') = const \cdot \int \exp\left(-i\vec{k}' \cdot \vec{r}'\right) V(\vec{r}'') \exp\left(i\vec{k} \cdot \vec{r}'\right) d^3r'$$

$$A_{-e,0}(\hat{n}') = const \cdot \int \exp\left(-i\vec{k}' \cdot \vec{r}'\right) V(\vec{r}') \exp\left(i\vec{k} \cdot \vec{r}'\right) d^3r'$$

Figure 7.7 X-ray beam scattered in the $\hat{n}'$ direction by electrons placed at $\vec{r}$ and at the origin **O**. The scattering amplitudes are given.

where the scattering amplitude $A(\hat{n}')$ is given by

$$A(\hat{n}') = const \int \exp\left(-i\vec{k}' \cdot \vec{r}'\right) V(\vec{r}') \exp\left(i\vec{k} \cdot \vec{r}'\right) d^3\vec{r}'. \quad (7.19)$$

The scattering amplitude $A(\hat{n}')$ completely determines the wave function $\psi_{\hat{n}'}(\vec{r})$.

Next, we will compare the scattering amplitude which corresponds to the X-ray beam scattered by an electron placed at $\vec{r}$ (see Fig. 7.7) with the scattering amplitude for the case of a scattering center placed at the origin **O**. The incoming X-ray beam propagates in the $\hat{n}$ direction and the scattering amplitudes for waves dispersed in the $\hat{n}'$ direction by electrons at $\vec{r}$ and at the origin **O** are $A_{-e,\vec{r}}(\hat{n}')$ and $A_{-e,0}(\hat{n}')$, respectively (see Fig. 7.7). Since $\vec{r}' = \vec{r} + \vec{r}''$, we have

$$A_{-e,\vec{r}}(\hat{n}') = const \int \exp\left[-i\vec{k}' \cdot (\vec{r} + \vec{r}'')\right] V(\vec{r}'') \exp\left[i\vec{k} \cdot (\vec{r} + \vec{r}'')\right] d^3r''$$

$$= \exp\left[i\left(\vec{k} - \vec{k}'\right) \cdot \vec{r}\right] \cdot A_{-e,0}(\hat{n}'), \quad (7.20)$$

as

$$A_{-e,0}\left(\hat{n}'\right) = \text{const} \int \exp\left(-i\vec{k}' \cdot \vec{r}''\right) V\left(\vec{r}''\right) \exp\left(i\vec{k} \cdot \vec{r}''\right) d^3r''.$$

(7.21)

So we finally obtain

$$A_{-e,\vec{r}}\left(\hat{n}'\right) = \exp\left[i\left(\vec{k} - \vec{k}'\right) \cdot \vec{r}\right] \cdot A_{-e,0}\left(\hat{n}'\right).$$

(7.22)

The last formula shows how the scattering amplitude corresponding to the wave scattered by the electron placed at $\vec{r}$ is expressed by the scattering amplitude of the wave scattered by an electron placed at the origin **O**. Below, we are going to rewrite this formula in a more convenient form.

The vector $\vec{s}$, defined in Fig. 7.8, represents a difference of unit vectors $\hat{n}$ and $\hat{n}'$ that are in the directions of wave vector $\vec{k}$ of the incident wave and of wave vector $\vec{k}'$ of the scattered wave, respectively. We have

$$\vec{k} - \vec{k}' = \frac{2\pi}{\lambda}\hat{n} - \frac{2\pi}{\lambda}\hat{n}' = \frac{2\pi}{\lambda}\left(\hat{n} - \hat{n}'\right) = \frac{2\pi}{\lambda}\vec{s}.$$

(7.23)

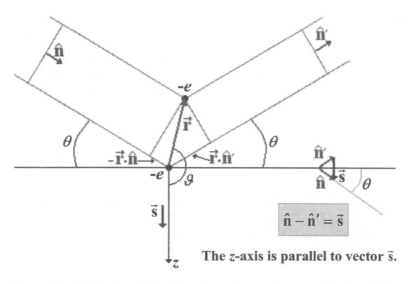

The z-axis is parallel to vector $\vec{s}$.

Figure 7.8 Scattering of the X-ray beam by an electron placed at $\vec{r}$. As a reference, the scattering by an electron placed at the origin is also considered. The z-axis is parallel to vector $\vec{s}$ defined in the draw.

Therefore, the scattering amplitude corresponding to the ray scattered by the electron at $\vec{r}$ can be expressed by the scattering amplitude corresponding to the ray scattered by an electron placed at the origin **O** as follows:

$$A_{-e,\vec{r}}\left(\hat{n}'\right) = A_{-e,0}\left(\hat{n}'\right)\exp\left[i\left(\vec{k} - \vec{k}'\right)\cdot\vec{r}\right]$$

$$= A_{-e,0}\left(\hat{n}'\right)\exp\left(i\frac{2\pi}{\lambda}\vec{r}\cdot\vec{s}\right). \qquad (7.24)$$

In the last formula, $\exp\left[i\left(\vec{k} - \vec{k}'\right)\cdot\vec{r}\right]$ represents a phase factor, which differentiates the wave scattered by an electron placed at $\vec{r}$, $\psi_{\hat{n}',\vec{r}}$, from the wave scattered by an electron placed at the origin **O**, $\psi_{\hat{n}',0}$, and $\left(\vec{k} - \vec{k}'\right)\cdot\vec{r} \equiv \Phi$ is the phase difference between the two waves. From Fig. 7.8, we can see that

$$\vec{r}\cdot\vec{s} = rs\cos\vartheta = 2r\sin\theta\cos\vartheta, \qquad (7.25)$$

since $s = 2\sin\theta$. So the phase difference Φ is given by

$$\Phi = \frac{2\pi}{\lambda}\vec{r}\cdot\vec{s} = \frac{4\pi}{\lambda}r\sin\theta\cos\vartheta \qquad (7.26)$$

and finally

$$A_{-e,\vec{r}}\left(\hat{n}'\right) = A_{-e,0}\left(\hat{n}'\right)\exp\left(i\Phi\right). \qquad (7.27)$$

Now we will return to our scattering center, that is to say, an atom composed of Z electrons. An atom represents an electron charge distributed in a certain volume with certain charge density distribution. We will designate the electron density with $\rho(\vec{r})$, then in an infinitesimal volume d^3r, shown in Fig. 7.9, there are $\rho(\vec{r})\,d^3r$ electrons that represent a charge equal to $-e\rho(\vec{r})\,d^3r$. At point $\vec{r}'$, the infinitesimal charge $-e\rho(\vec{r})\,d^3r$ produces an electrostatic potential $dV\left(\vec{r}''\right)$ which scatters an incident X-ray wave. The scattering amplitude corresponding to the wave scattered by charge $-e\rho(\vec{r})\,d^3r$ (see Fig. 7.10) will be proportional to it, since the electrostatic potential which determines the scattering amplitude is proportional to this charge, and therefore

$$dA_{\rho,\vec{r}}\left(\hat{n}'\right) \propto \left[-e\rho(\vec{r})\,d^3r\right]. \qquad (7.28)$$

The scattering amplitude $dA_{\rho,\vec{r}}\left(\hat{n}'\right)$ differs from the scattering amplitude for a wave dispersed by an electron at origin **O**, $A_{-e,0}\left(\hat{n}'\right)$,

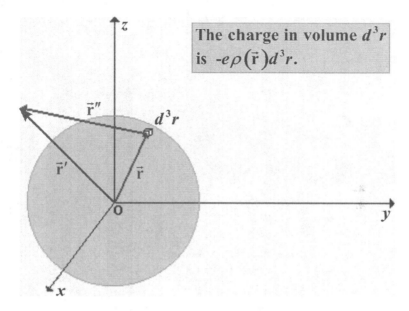

The charge in volume d^3r is $-e\rho(\vec{r})d^3r$.

Figure 7.9 An isolated spherical atom with electron density $\rho(\vec{r})$. The infinitesimal electron charge within the volume d^3r at a point $\vec{r}$ produces an infinitesimal electrostatic potential $dV(\vec{r}'')$ at a point $\vec{r}'$.

by two terms, $\rho(\vec{r})\,d^3r$ and $\exp(i\Phi)$. The first term represents the fraction of electrons that are in the d^3r volume at point $\vec{r}$, while the second term (phase factor) differentiates the wave scattered by the charge at $\vec{r}$ from the wave scattered by the electron placed at the origin **O**. Therefore

$$dA_{\rho,\vec{r}}(\hat{n}') = A_{-e,0}(\hat{n})\,\rho(\vec{r})\,d^3r\exp(i\Phi) \qquad (7.29)$$

and the scattering amplitude corresponding to the wave scattered by Z electrons of an atom is given by

$$A_\rho(\hat{n}') = A_{-e,0}(\hat{n}')\int\rho(\vec{r})\exp(i\Phi)d^3r. \qquad (7.30)$$

The phase difference Φ, defined by Eq. 7.26, contains an angle ϑ, which is specified in Fig. 7.11. In the case of spherical symmetry, when the spherically averaged atomic electron density is assumed, we have

$$\rho(\vec{r}) = \rho(r) \qquad (7.31)$$

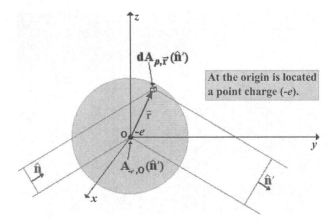

Figure 7.10 X-ray scattered elastically in the $\hat{n}'$ direction by an infinitesimal charge $-e\rho(\vec{r})d^3r$ placed at $\vec{r}$. As a reference, the scattering by an electron placed at the origin **O** is also considered.

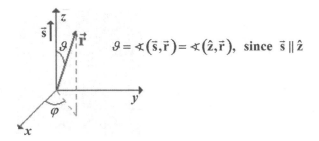

Figure 7.11 Definition of the spherical coordinates r, ϑ, and φ with respect to vector $\vec{s}$ defined in Fig. 7.8.

and then

$$A_\rho(\hat{n}') = A_{-e,0}(\hat{n}') \int \rho(r) \exp\left(i\frac{4\pi}{\lambda}r\sin\theta\cos\vartheta\right) d^3r$$

$$= A_{-e,0}(\hat{n}') \int_0^{2\pi} d\phi \int_0^{\pi} d\vartheta \int_0^{\infty} \rho(r)\exp\left(iar\cos\vartheta\right) r^2\sin\vartheta\,dr,$$

$$(7.32)$$

where in the second line, we have expressed $d^3r = r^2\sin\vartheta\,dr\,d\vartheta\,d\phi$ in spherical coordinates, and we have proposed the following abbreviation within the exponential function

$$\frac{4\pi}{\lambda}\sin\theta \equiv a. \tag{7.33}$$

Integrating for ϕ and ϑ in Eq. 7.32, the following result is obtained:

$$\int_0^{2\pi} d\phi \int_0^\pi \exp\left(iar\cos\vartheta\right)\sin\vartheta\, d\vartheta = 2\pi \int_{-1}^1 \exp\left(iarx\right) dx$$

$$= \frac{2\pi}{iar}\left[\exp\left(iar\right) - \exp\left(-iar\right)\right] = \frac{4\pi}{ar}\sin\left(ar\right), \tag{7.34}$$

where integrating for ϑ a variable change, $\cos\vartheta = x$, was made. Lastly, we have

$$A_\rho\left(\hat{n}'\right) = \frac{4\pi}{a} A_{-e,0}\left(\hat{n}'\right) \int_0^\infty \rho(r)r\sin\left(ar\right) dr. \tag{7.35}$$

Now we will define the *atomic scattering factor* f (for a given scatter direction) as the ratio of the scattering amplitude corresponding to the wave scattered by all the electrons of the atom and the scattering amplitude of the wave scattered by one electron placed at the center of the atom:

$$f = \frac{A_\rho\left(\hat{n}'\right)}{A_{-e,0}\left(\hat{n}'\right)} = \frac{4\pi}{a}\int_0^\infty \rho(r)r\sin\left(ar\right) dr, \quad \text{where} \quad a = \frac{4\pi}{\lambda}\sin\theta.$$

$$\tag{7.36}$$

The atomic scattering factor is also called the *atomic form factor*.

Until now, we have introduced formulas that describe the constructive interference of X-rays considering the case of scattering by equal atoms of a crystal structure with one-atom basis. Now, we will analyze the case when a crystal structure has p atoms per unit cell. The position vectors of the p atoms are

$$\vec{r}_m = u_m\vec{a}_1 + v_m\vec{a}_2 + w_m\vec{a}_3, \quad \text{where} \quad m = 1, 2, \ldots, p. \tag{7.37}$$

The vectors $\vec{a}_1$, $\vec{a}_2$, and $\vec{a}_3$ define the unit cell and u_m, v_m, w_m are fractional coordinates of the basis atoms. Figure 7.12 shows a two-dimensional crystal structure with two atoms in the unit cell. We can see that a structure which has the basis composed of p atoms may be considered as a superposition of p substructures of the same type with one atom basis each. Since all the substructures are of the same type, the constructive interference for each substructure is described by the same Laue conditions:

$$\vec{a}_i \cdot \left(\hat{n}' - \hat{n}\right) = g_i\lambda, \quad \text{where } i = 1, 2, 3 \text{ and } g_i \in \mathbb{Z}. \tag{7.38}$$

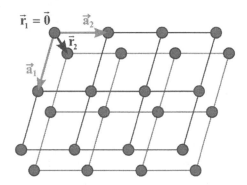

Figure 7.12 A two-dimensional crystal structure. Basis vectors $\vec{a}_1$ and $\vec{a}_2$ define the unit cell. The atomic basis is composed of two atoms at positions $\vec{r}_1$ and $\vec{r}_2$.

7.5 Structure Factor

Now we will analyze the conditions in which the constructive interference occurs because of the scattering of X-rays by atoms belonging to some crystal structure. Figure 7.13a shows an example of a conventional unit cell of a two-dimensional crystal structure. This cell contains two atoms, one at position given by $\vec{r}_m$ and the other, of a different type, at the origin. It is easy to show that the scattering amplitude corresponding to the wave scattered by an atom m placed at $\vec{r}_m$ may be expressed by the scattering amplitude corresponding to the same wave scattered by an atom m but placed at the origin (see Fig. 7.13b). The scattering amplitude $A^m_{\rho, \vec{r}_m}\left(\hat{n}'\right)$ is expressed by $A^m_{\rho, 0}\left(\hat{n}'\right)$ as follows:

$$
\begin{aligned}
A^m_{\rho, \vec{r}_m}\left(\hat{n}'\right) &= A^m_{\rho, 0}\left(\hat{n}'\right) \exp\left[i\left(\vec{k} - \vec{k}'\right) \cdot \vec{r}_m\right] \\
&= A^m_{\rho, 0}\left(\hat{n}'\right) \exp\left(i\,\Phi_m\right),
\end{aligned}
$$

$$
\text{where } \Phi_m = \frac{2\pi}{\lambda}\vec{r}_m \cdot \vec{s}. \tag{7.39}
$$

In three dimensions, the X-ray scattering in the $\hat{n}'$ direction corresponds to the mirror reflection from (hkl) planes orthogonal to vector $\vec{s}$, which is parallel to the $\vec{b}_{hkl}$ vector. Let us transform the

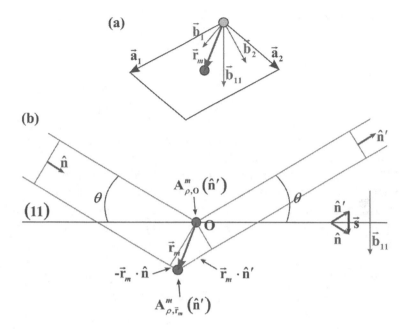

Figure 7.13 (a) Conventional unit cell of a two-dimensional crystal structure defined by lattice vectors $\vec{a}_1$ and $\vec{a}_2$. The reciprocal lattice is defined by vectors $\vec{b}_1, \vec{b}_2$. The unit cell contains two atoms of different types. (b) X-ray scattering by the atom at position $\vec{r}_m$ in (a) and the same atom placed at the origin. The vector $\vec{b}_{hk} = \vec{b}_{11} = \vec{b}_1 + \vec{b}_2$ is orthogonal to the family of (11) planes.

expression for the phase factor Φ_m as follows:

$$
\begin{aligned}
\Phi_m &= \left(\vec{k} - \vec{k}'\right) \cdot \vec{r}_m = \frac{2\pi}{\lambda}\left(\hat{n} - \hat{n}'\right) \cdot \left(u_m \vec{a}_1 + v_m \vec{a}_2 + w_m \vec{a}_3\right) \\
&= \frac{2\pi}{\lambda}\left[u_m \left(\hat{n} - \hat{n}'\right) \cdot \vec{a}_1 + v_m \left(\hat{n} - \hat{n}'\right) \cdot \vec{a}_2 + w_m \left(\hat{n} - \hat{n}'\right) \cdot \vec{a}_3\right] \\
&= 2\pi \left(nhu_m + nkv_m + nlw_m\right),
\end{aligned}
\tag{7.40}
$$

where in the last step we made use of the following:

$$
\begin{cases}
\left(\hat{n} - \hat{n}'\right) \cdot \vec{a}_1 = nh\lambda \\
\left(\hat{n} - \hat{n}'\right) \cdot \vec{a}_2 = nk\lambda \\
\left(\hat{n} - \hat{n}'\right) \cdot \vec{a}_3 = nl\lambda
\end{cases}
\tag{7.41}
$$

Equations 7.41 represent the Laue conditions for the family of (hkl) planes. In the case of Φ_m a convention different from the one proposed before for the *hkl* indices will be adopted, namely

$$\begin{cases} nh \equiv h \\ nk \equiv k \ , \\ nl \equiv l \end{cases} \qquad (7.42)$$

which in practice implies that the planes will be indexed by the *hkl* indices given by the smallest integer numbers (as we have done up to now) and their multiples. For instance, according to this convention the sets of planes (100), (200), (300), ... will be good. Of course, only (100) represents a family of real lattice planes and the planes (200), (300), ... are fictitious. In fact, the *hkl* indices which are multiples of the smallest integer numbers correspond to reflections from a real family of planes, but represent reflections of higher order (*n*-order reflections). In the case of the 100 reflection the difference in lengths of optical paths of rays reflected from two consecutive (100) planes is 1λ, while in the (200) case this difference is 2λ and occurs from the same (100) planes. Finally, the scattering amplitude $A^m_{\rho,\vec{r}_m}(\hat{n}')$ can be expressed by $A^m_{\rho,0}(\hat{n}')$ in the following way:

$$A^m_{\rho,\vec{r}_m}(\hat{n}') = A^m_{\rho,0}(\hat{n}') \exp\left[i 2\pi \left(h u_m + k v_m + l w_m\right)\right] \qquad (7.43)$$

and the scattering amplitude corresponding to the wave scattered by the set of *p* atoms belonging to the conventional unit cell is given by the expression

$$A_p(\hat{n}') = \sum_{m=1}^{p} A^m_{\rho,\vec{r}_m}(\hat{n}')$$

$$= \sum_{m=1}^{p} A^m_{\rho,0}(\hat{n}') \exp\left[i 2\pi \left(h u_m + k v_m + l w_m\right)\right]. \qquad (7.44)$$

$A_p(\hat{n}')$ expresses the scattering amplitude corresponding to the wave scattered in the $\hat{n}'$ direction by all *p* substructures in the maximum of interference of X-rays diffracted from their (*hkl*) planes. We will now introduce a new parameter, the *structure factor* $F(hkl)$, defined as a ratio of $A_p(\hat{n}')$ and the scattering amplitude corresponding to the wave scattered by a single electron in the same $\hat{n}'$ direction

$$F(hkl) = \frac{1}{A_{-e,0}(\hat{n}')} \sum_{m=1}^{p} A^m_{\rho,0}(\hat{n}') \exp\left[i 2\pi \left(h u_m + k v_m + l w_m\right)\right]$$

$$= \sum_{m=1}^{p} \frac{A^m_{\rho,0}(\hat{n}')}{A_{-e,0}(\hat{n}')} \exp\left[i 2\pi \left(h u_m + k v_m + l w_m\right)\right], \qquad (7.45)$$

finally

$$F(hkl) = \sum_{m=1}^{p} f_m \exp\left[i2\pi\left(hu_m + kv_m + lw_m\right)\right]. \qquad (7.46)$$

If all the atoms in the crystal are of the same type, we have

$$f_1 = f_2 = \ldots\ldots = f_p = f \qquad (7.47)$$

and

$$F(hkl) = f \sum_{m=1}^{p} \exp\left[i2\pi\left(hu_m + kv_m + lw_m\right)\right] = f \cdot S(hkl), \quad (7.48)$$

where

$$S(hkl) = \sum_{m=1}^{p} \exp\left[i2\pi\left(hu_m + kv_m + lw_m\right)\right]. \qquad (7.49)$$

$S(hkl)$ is called the *geometric structure factor*. The structure factor $F(hkl)$ is in general a complex number. The diffraction intensity is proportional to $|F(hkl)|^2$.

We will go now to the example of the *bcc* crystal structure. Its conventional unit cell is the cubic *I* cell shown in Fig. 7.14. It contains two atoms of coordinate triplets $0, 0, 0$ and $\frac{1}{2}, \frac{1}{2}, \frac{1}{2}$. In this case we have

$$\begin{aligned} S(hkl) &= \sum_{m=1}^{2} \exp\left[i2\pi\left(hu_m + kv_m + lw_m\right)\right] \\ &= 1 + \exp\left[i\pi\left(h + k + l\right)\right], \end{aligned} \qquad (7.50)$$

and therefore

$$S(hkl) = \begin{cases} 0, & \text{if } (h+k+l) \text{ is odd} \\ 2, & \text{if } (h+k+l) \text{ is even} \end{cases}. \qquad (7.51)$$

Thus, in the case of the *bcc* crystal structure, the interference maximums 100, 111, 210, 300, ... will not be observed. Moreover, the interference maximum 100 is not observed, while the maximum 200 is observed. In order to understand the last comment it is necessary to realize that the family of (100) planes does not contain the planes that are defined by centering points, while the family of (200) planes contains them. So, the interference maximum 100 is destroyed by the presence of such planes in the *bcc* crystal structure [but not included in the family of (100) planes]. We could learn from this example that in the case of a centered unit cell certain conditions apply to the reflection *hkl*, the so-called *integral reflection conditions*, given in Table 7.1. The interference maximum is not observed if the reflection condition is not satisfied.

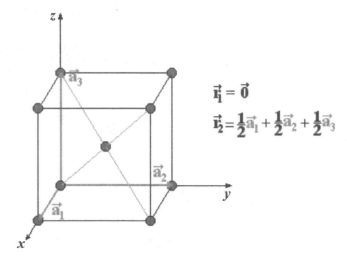

Figure 7.14 Conventional unit cell cubic I for the bcc structure. This cell contains two atoms located at $\vec{r}_1$ and $\vec{r}_2$.

Table 7.1 Integral reflection conditions for centering cells

Centering type of cell	Reflection condition
Primitive	None
C-face centered	$h + k = 2n$
A-face centered	$k + l = 2n$
B-face centered	$h + l = 2n$
Body centered	$h + k + l = 2n$
All-face centered	h, k, l all odd or all even

Problems

Exercise 7.1 Find the expression for the square of the modulus of the structure factor, $|F(hkl)|^2$, for the CsCl crystal structure and reduce this expression for the case when the two atoms within the conventional unit cell are of the same type (the case of the bcc structure).

Exercise 7.2 The geometric structure factor $S(hkl)$ is given by Eq. 7.49.

(a) Find the expression of S for the *fcc* structure as a function of the *hkl* indices.
(b) Calculate the values of the geometric structure factors S for the following families of planes: (100), (110), (111), (211), and (422).
(c) Give the values for the geometric structure factor S for the *fcc* structure.

Exercise 7.3 Show that the geometric structure factor S_{diamond} for the diamond structure is a product of the geometric structure factor S_{fcc} for the *fcc* structure and the factor $\left\{1 + \exp\left[i\frac{\pi}{2}(h+k+l)\right]\right\}$. When does S_{diamond} have a real value?

Exercise 7.4 Find the expression for the atomic scattering factor f defined by Eq. 7.36 when the X-ray beam is parallel to a lattice plane ($\theta = 0$).

Appendix

A.1 Solutions to Exercises of Chapter 1

Exercise 1.1 Since the structure shown in Fig. 1.29 does not possess reflection points, the origin of the unit cell is arbitrary. A possible choice for the lattice points is shown in Fig. A.1. The cell in this figure is conventional.

Figure A.1 Possible choice of lattice points for the structure from Fig. 1.29.

Exercise 1.2 The three structures shown in Fig. 1.1 possess reflection points, so the line group of each of them is p-m, whereas the structure from Fig. 1.29 does not possess reflection points and its line group is p1.

Exercise 1.3 The structure shown in Fig. 1.13 does not possess any symmetry element different from identity, so the origin of its unit cell may be arbitrary. It usually coincides with the center of one of the atoms conforming the atomic basis. Figure A.2 shows another choice for the origin of the conventional unit cell defined by basis vectors $\vec{a}$, $\vec{b}$ in Fig. 1.13.

Exercise 1.4 By placing an additional point in the geometric center of each equilateral triangle of the lattice from Fig. 1.30, we obtain another hexagonal lattice defined by basis vectors $\vec{a}'$, $\vec{b}'$ shown in Fig. A.3.

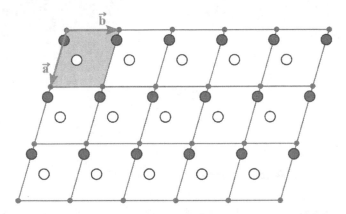

Figure A.2 A choice for the origin of a conventional unit cell for the two-dimensional structure from Fig. 1.13.

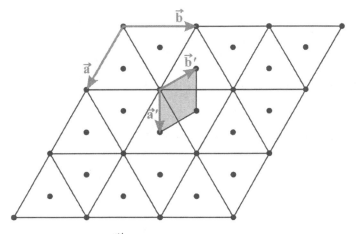

Figure A.3 Vectors $\vec{a}'$, $\vec{b}'$ are basis vectors of a new hexagonal lattice obtained by placing an additional point in the geometric center of each equilateral triangle of the lattice shown in Fig. 1.30.

Exercise 1.5

Ad (a) The twofold rotation points in a hexagonal lattice are specified in Fig. A.4a.

Ad (b) The location of twofold rotation points of the centered rectangular lattice is shown in Fig. A.4b.

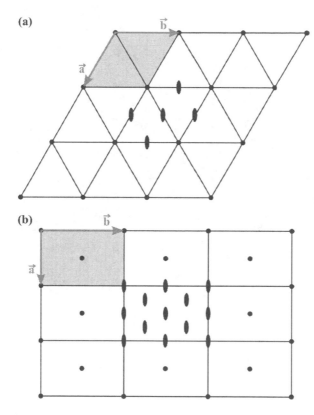

Figure A.4 (a) Different twofold rotation points within the primitive unit cell of the hexagonal lattice. (b) Locations of twofold rotation points in the centered rectangular lattice.

Exercise 1.6

Ad (a) If we attach to each lattice point of a hexagonal lattice a basis that has two identical atoms in the positions given by vectors $\vec{r}_1 = 2/3\vec{a} + 1/3\vec{b}$ and $\vec{r}_2 = 1/3\vec{a} + 2/3\vec{b}$, then the honeycomb structure shown in Fig. A.5a will be obtained.

Ad (b) In the two-dimensional boron nitride structure, shown in Fig. A.5b, only threefold rotation points are present.

Exercise 1.7 The hexagonal lattice from Fig. 1.30 with an atomic basis that has a large atom in the position given by vector $\vec{r}_1 = \vec{0}$ and two small atoms in the positions given by vectors $\vec{r}_2 = 2/3\vec{a} + 1/3\vec{b}$

(a)

(b)

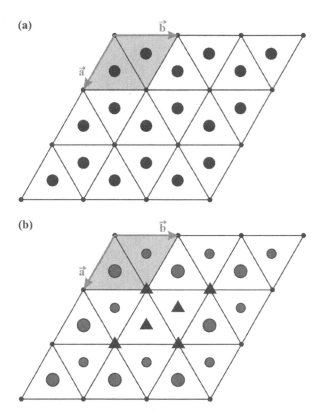

Figure A.5 (a) Honeycomb structure. (b) Locations of threefold rotation points in the two-dimensional boron nitride structure.

and $\vec{r}_3 = 1/3\vec{a} + 2/3\vec{b}$ is depicted in Fig. A.6. This structure is a superposition of a hexagonal structure built of larger atoms and a honeycomb structure built of smaller atoms. In Fig. A.6, we can find easily the hexagon from Fig. 1.32a.

Exercise 1.8 In Fig. A.7, we can see a honeycomb structure obtained from a superposition of two identical hexagonal substructures.

Exercise 1.9

Ad (a) The lattice is hexagonal.
Ad (b) The primitive unit cell for this lattice is shown in Fig. A.8.

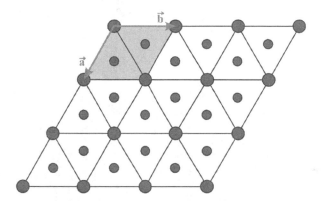

Figure A.6 A two-dimensional crystal structure that is a superposition of two substructures: hexagonal and honeycomb.

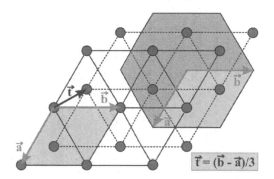

Figure A.7 The honeycomb structure seen as a superposition of two identical hexagonal substructures shifted one with respect to the other by vector $\vec{t} = \left(\vec{b} - \vec{a}\right)/3$, where $\vec{a}$, $\vec{b}$ are the basis vectors of the hexagonal substructures and also of the resulting honeycomb structure.

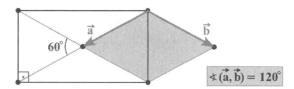

Figure A.8 Hexagonal lattice. Vectors $\vec{a}$, $\vec{b}$ define the primitive unit cell for this lattice.

Exercise 1.10 The lattice directions in a square lattice denoted by [1$\bar{1}$] and [11] are drawn in Fig. A.9a. In Fig. A.9b are depicted the reflection lines represented by those directions.

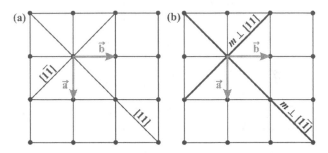

Figure A.9 Square lattice. In (a) are plotted lattice directions denoted by [1$\bar{1}$] and [11], whereas in (b) reflection lines $m\perp$[1$\bar{1}$] and $m\perp$[11].

Exercise 1.11

Ad (a) The hexagonal lattice directions [10], [01], and [$\bar{1}\bar{1}$] and the reflection lines represented by them are shown in Figs. A.10a and A.10b, respectively.

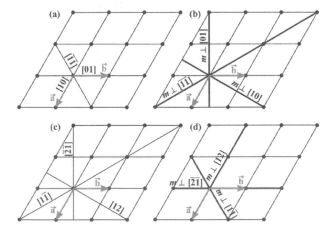

Figure A.10 Hexagonal lattice. In (a) are shown lattice directions denoted by [10], [01], and [$\bar{1}\bar{1}$], whereas in (b) reflection lines $m\perp$[10], $m\perp$[01], and $m\perp$[$\bar{1}\bar{1}$]. (c) Lattice directions denoted by [1$\bar{1}$], [12], and [$\bar{2}\bar{1}$]. (d) Reflection line orthogonal to lattice directions from (c).

Ad (b) Similar information to that given in (a), but now for lattice directions [1$\bar{1}$], [12], and [$\bar{2}\bar{1}$] is given in Figs. A.10c and A.10d.

Exercise 1.12

Ad (a) The structure defined in Fig. 1.32b and its hexagonal lattice are drawn in Fig. A.11.

Ad (b) In this figure are also shown the symmetry elements indicated in the Hermann–Mauguin symbol *p3m*1.

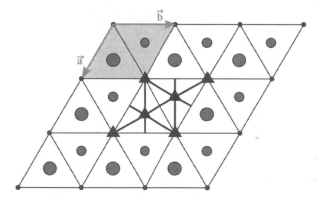

Figure A.11 The relative locations of the symmetry elements included in the Hermann–Mauguin symbol for the plane group of the α-BN structure, which is *p3m*1.

Exercise 1.13 The structure defined in Fig. 1.32c, its hexagonal lattice, and the symmetry elements indicated in the Hermann–Mauguin symbol *p6mm* are shown in Fig. A.12.

Exercise 1.14

Ad (a) The diagram of symmetry elements of the rectangular lattice plane group is drawn in Fig. A.13a.

Ad (b) The same as in (a) but for the centered rectangular lattice is given in Fig. A.13b.

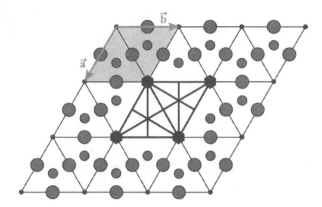

Figure A.12 A two-dimensional structure defined in Fig. 1.32c. Shown are the relative locations of the symmetry elements indicated in the Hermann–Mauguin symbol for the plane group of this structure, which is *p6mm*.

Exercise 1.15

Ad (a) Point symmetry elements of a rectangular lattice are shown in Fig. A.14a. The Hermann–Mauguin symbol for the point group of this lattice is *2mm*.

Ad (b) The centered rectangular lattice has the same point symmetry as the rectangular lattice. This is shown in Fig. A.14b.

Exercise 1.16 Six boron and two carbon atoms belong to the unit cell of the two-dimensional B_3C structure shown in Fig. 1.34. The 4 boron atoms placed at the vertices of the cell contribute to the cell with one atom, and the 4 boron atoms placed on the edges of the cell contribute with 2 boron atoms (each of them with a half of atom). There are still 3 entire boron atoms within the cell what gives a total of 6 boron atoms. In a similar way, the number of carbon atoms can be calculated.

Exercise 1.17

Ad (a) The cell origin is on *m* (see Fig. A.15).

Ad (b) The Hermann–Mauguin symbol of the plane group of the two-dimensional B_3C structure shown in Fig. 1.34 is *p1m1*. The diagram of symmetry elements of the plane group *p1m1* is shown in Fig. A.15.

(a)

(b)

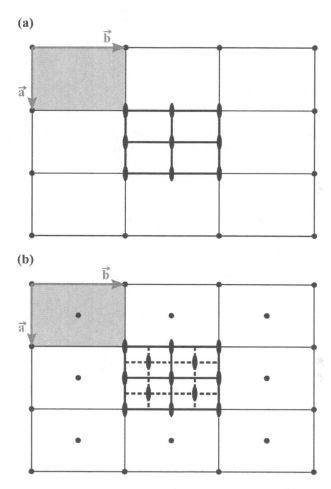

Figure A.13 (a) Diagram of symmetry elements of the plane group of a rectangular lattice. The Hermann–Mauguin symbol of this plane group is *p2mm*. (b) Diagram of symmetry elements of the plane group of a centered rectangular lattice. The Hermann–Mauguin symbol of the plane group is *c2mm*. The glide lines, shown in the diagram, are within the symmetry elements of the *c2mm* plane group but are not included in the Hermann–Mauguin symbol.

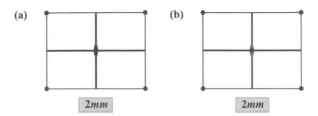

Figure A.14 (a) Point symmetry elements of a rectangular lattice. (b) Point symmetry elements of a centered rectangular lattice.

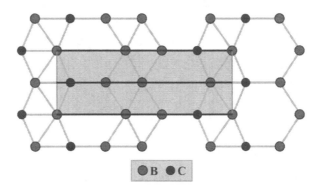

Figure A.15 Diagram of symmetry elements of the *p1m1* plane group of the two-dimensional B$_3$C structure shown in the figure. The basis vectors $\vec{a}$, $\vec{b}$ are defined in Fig. 1.34.

Exercise 1.18

Ad (a) The diagram of symmetry elements for the plane group of the square lattice is shown in Fig. A.16.

Ad (b) In the Hermann–Mauguin symbol *p4mm* there are not indicated twofold rotation points in the middles of square edges and glide lines orthogonal to lattice directions [1$\bar{1}$] and [11] (see Fig. A.16).

Exercise 1.19

Ad (a) A conventional unit cell of the NaCl(001) surface is shown in Fig. A.17. The origin of the conventional unit cell could be also chosen at the center of the Na ion.

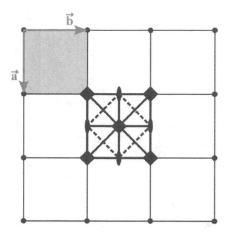

Figure A.16 Diagram of symmetry elements of the plane group of the square lattice. Its Hermann–Mauguin symbol is *p4mm*.

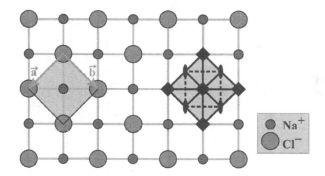

Figure A.17 Ideal (001) surface of NaCl. Shown is the diagram of symmetry elements of its plane group *p4mm*.

Ad (b) The cell origin is at *4mm* and the Hermann–Mauguin symbol of the plane group is *p4mm*. The diagram of symmetry elements is shown in Fig. A.17.

Exercise 1.20 The glide reflection of an atom through a glide line *b* in the two-dimensional boron structure depicted in Fig. 1.36 is illustrated in Fig. A.18.

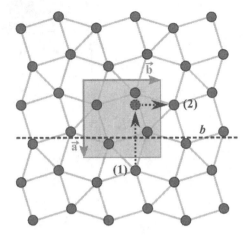

Figure A.18 Demonstration of a glide reflection of the atom labeled as (1) to a position labeled as (2). This glide reflection consists in reflection through the line *b* and the translation by vector $1/2\vec{b}$.

Exercise 1.21 The plane group of the boron structure shown in Fig. 1.36 is *p4gm*. Its diagram of symmetry elements is shown in Fig. A.19. The symmetry elements that are not indicated in the Hermann–Mauguin symbol are twofold rotation points in the middle

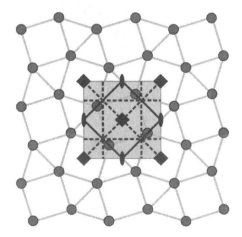

Figure A.19 Diagram of symmetry elements of the *p4gm* plane group of the two-dimensional boron structure shown in the figure. The basis vectors $\vec{a}$, $\vec{b}$ are defined in Fig. 1.36.

of cell edges and glide lines orthogonal to lattice directions $[1\bar{1}]$ and $[11]$ (see Fig. A.19). The cell origin is at $41g$.

Exercise 1.22

Ad (a) The atomic basis is formed by 3 boron atoms.

Ad (b) The plane group of the boron structure shown in Fig. 1.37 is *p2mm*. Its diagram of symmetry elements is shown in Fig. A.20.

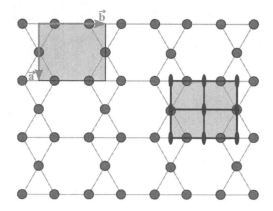

Figure A.20 Diagram of symmetry elements of the *p2mm* plane group of the two-dimensional boron structure shown in the figure.

Ad (c) The cell origin is at *2mm*. We can see in Fig. A.20 that a conventional origin can be placed also at points $0, \frac{1}{2}; \frac{1}{2}, 0;$ and $\frac{1}{2}, \frac{1}{2}$ (all of them lying at *2mm*).

Exercise 1.23 The labels of carbon atoms in different Wyckoff positions, specified in Table 1.4, are indicated in Fig. A.21a.

Exercise 1.24 The boron atoms in different Wyckoff positions, specified in Table 1.4, are highlighted by labels in Fig. A.21b.

Exercise 1.25

Ad (a) The plane group of the two-dimensional B_3C structure shown in Fig. 1.34 is *p1m1* (see Fig. A.15).

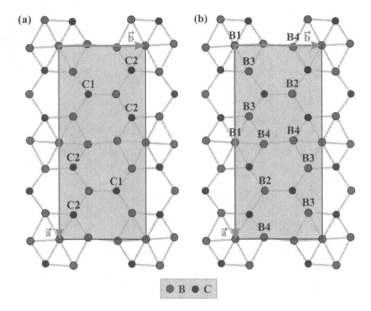

B ● C

Figure A.21 Two-dimensional B_2C structure. (a) The carbon atoms C1 belong to the Wyckoff special positions $2c$ and the atoms C2 belong to the Wyckoff special positions $4d$. (b) The boron atoms B1 are in Wyckoff special position $2a$, B2 in special position $2c$, and B3 and B4 in special position $4d$.

Ad (b) The Wyckoff positions for B and C atoms that lie within the unit cell are listed in Table A.1 (see Fig. 1.27). The atoms that are listed in the table are identified in Fig. A.22.

Table A.1 Wyckoff positions and coordinates of B and C atoms that lie within the unit cell of the two-dimensional B_3C structure shown in Fig. 1.34

Atom	Wyckoff position	Coordinates of equivalent positions
B1	$1a$	$0, 0$
B2	$1a$	$0, 0.390$
B3	$1a$	$0, 0.602$
B4	$1b$	$\frac{1}{2}, 0.273$
B5	$1b$	$\frac{1}{2}, 0.484$
B6	$1b$	$\frac{1}{2}, 0.874$
C1	$1a$	$0, 0.795$
C2	$1b$	$\frac{1}{2}, 0.080$

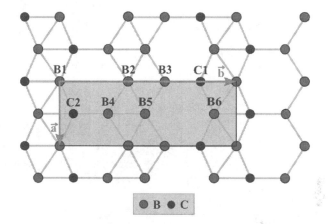

Figure A.22 Two-dimensional B_3C structure. The highlighted boron and carbon atoms correspond to Wyckoff positions specified in Table A.1.

Exercise 1.26

Ad (a) The conventional unit cell of the NaCl(001) surface, with the origin at the center of Cl^-, is shown in Fig. A.17.

Ad (b) In the case of the *p4mm* plane group, the site symmetry for the Wyckoff special position 1*a* is 4*mm* as the point of coordinates $0, 0$ coincides with a fourfold rotation point and at the same time with the common point of the following reflection lines: $m \perp [10]$, $m \perp [01]$, $m \perp [1\bar{1}]$, and $m \perp [11]$ (see Fig. A.17).

Ad (c) The site symmetry for the Wyckoff special position 1*b* (plane group *p4mm*) is also 4*mm*; see the explanation given in (b).

Exercise 1.27

Ad (a) The basis atoms with their coordinates are shown in Fig. A.23.

Ad (b) Taking $x = 0.173$ and adding 1 in the case of a negative coordinate, we obtain the coordinates reported in Table 1.6.

Ad (c) In the case of the plane group *p4gm* the site symmetry for the Wyckoff special position 4*c* is ..*m* (see Fig. A.19). The position of the first dot indicates that there is no rotation

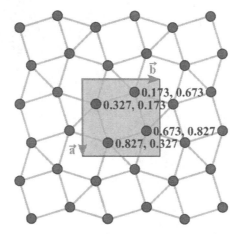

Figure A.23 Two-dimensional boron structure. The coordinates of the basis atoms are given (in units of a).

around a fourfold rotation point within the site symmetry point-group elements. The second dot indicates the absence of reflection trough lines orthogonal to directions [10] or [01] (the basis vectors $\vec{a}$, $\vec{b}$ are defined in Fig. 1.36 or A.18).

Exercise 1.28

Ad (a) Atoms B1 and B2 are identified in Fig. A.24.

Ad (b) The site symmetry of the special position $1c$ is $2mm$ and of the special position $2g$ is $.m.$ (see Fig. A.20).

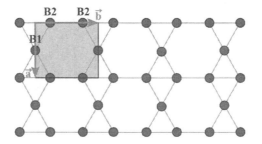

Figure A.24 In the two-dimensional boron structure shown in the figure, the atom B1 belongs to the Wyckoff special position $1c$ and atoms B2 belong to the Wyckoff special position $2g$ (plane group $p2mm$).

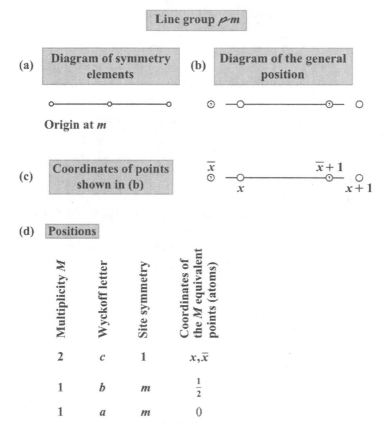

Figure A.25 The Pm line group. (a) Diagram of symmetry elements. (b) Diagram of the general position. (c) Coordinates of points shown in (b). (d) Wyckoff positions.

Exercise 1.29 Wyckoff general and special positions of the line group Pm are given in Fig. A.25.

Exercise 1.30

Ad (a) The line group of the structure from Fig. 1.38 is Pm.
Ad (b) Atom A is in the Wyckoff special position $1a$ and the two atoms B are in general position $2c$ (see Fig. A.25).

A.2 Solutions to Exercises of Chapter 2

Exercise 2.1 There are two more twofold rotation axes in the case of a square prism. For each of them, 4 of the 8 vertices of the prism define two segments orthogonal to the rotation axis, while the rest of the vertices define a rectangle whose plane is orthogonal to the axis, and the axis crosses the segments and the rectangle in their geometric centers.

Exercise 2.2

Ad (a) & (b) In order to draw each rotation axis, locate first the positions of two points that define it. If necessary find these points graphically. For instance, in the tetrahedron, the threefold rotation axes connect vertices with the centers of the opposite faces.

Exercise 2.3 We will show that the hexagonal faces of the truncated octahedron from Fig. 2.46 are regular. For this purpose an auxiliary Fig. A.26 is drawn. This figure shows a vertical cross section of the 4 solid figures depicted in Fig. 2.46. The cross sections of the cubes (the squares of side $(3/2)\,a$ and a) are parallel to their front faces

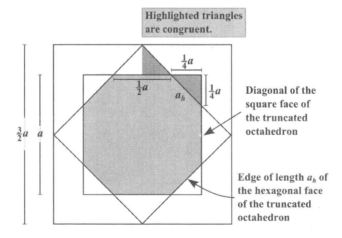

Figure A.26 A vertical cross section of the 4 solid figures depicted in Fig. 2.46.

and contain the common geometric center of the 4 solid figures. The third square in Fig. A.26 [inscribed in the square of edge $(3/2)\,a$] corresponds to the cross section of the regular octahedron, and the highlighted octagon represents the cross section of the truncated octahedron. The shortest edges of the octagon (of length a_h) coincide with 3 edges of the hexagonal faces shown in Fig. 2.46. The other 4 edges of the octagon [of length $(1/2)\,a$] coincide with the diagonals of the square faces of the truncated octahedron. The hexagonal faces have 3 common edges with the squares, of length a_s, shown in Fig. 2.46. We will show now that $a_h = a_s$, so the hexagons are regular. From Fig. A.26 we can see that

$$a_h = \sqrt{\left(\frac{a}{4}\right)^2 + \left(\frac{a}{4}\right)^2} = \frac{\sqrt{2}}{4}a, \tag{A.1}$$

while the edge length of the square face fulfills the relation $2a_s^2 = [(1/2)\,a]^2$, which gives

$$a_s = \frac{\sqrt{2}}{4}a. \tag{A.2}$$

From Eqs. A.1 and A.2 we finally get $a_h = a_s$, so the hexagonal face of the truncated octahedron is just a regular hexagon of edge length $\left(\sqrt{2}/4\right)a$.

Exercise 2.4

Ad (a) & (b) The rhombic dodecahedron has 14 vertices. Each of the 4 threefold or 3 fourfold axes of the rhombic dodecahedron (7 in total) is defined by two of its vertices. To draw each axis, we may substitute one of the vertices by the geometric center of the dodecahedron.

Exercise 2.5 The set of 27 points from Fig. 2.48 has 4 threefold and 3 fourfold rotation axes. A fourfold (threefold) rotation axis generates from one axis 3 (2) other threefold (fourfold) axes. In order to demonstrate the existence of a threefold or fourfold rotation axis, follow the discussion performed in Section 2.4 for the case of the set of 14 points. Choose one axis of each type and draw plane figures (defined by points of the set) in planes orthogonal to the axis. For instance, a regular hexagon shown in Fig. 2.48 is lying in a plane orthogonal to a threefold rotation axis. There are 7 plane figures (among them the regular hexagon and two small triangles that are inscribed in larger ones) defined by 24 points, which are lying in 5 planes orthogonal to this axis. The axis crosses each plane figure in its geometric center.

Exercise 2.6 The set of 27 points defined in Fig. 2.48 has 6 twofold rotation axes. For each axis, three points of this set are placed on it and the rest 24 define segments and plane figures (rectangles) that are orthogonal to this axis, which crosses the geometric center of each segment and plane figure.

Exercise 2.7 In order to show that the A-face centered and C-face centered monoclinic lattices are equivalent in the case of the b-axis setting, find a body centered monoclinic cell for each of the lattices shown in Fig. 2.49.

Exercise 2.8 There are 6 NNs, 6 NNNs, 6 TNNs, and 12 fourth nearest neighbors of a lattice point in a two-dimensional hexagonal lattice at distances a, $a\sqrt{3}$, $a\sqrt{4}$, and $a\sqrt{7}$ from this point, respectively. The number of neighbors of each order has to be a multiple of 6, because the lattice points in a two-dimensional hexagonal lattice coincide with sixfold rotation points.

Exercise 2.9

Ad (a) When the lengths of the edges a, b of the conventional rectangular unit cell of a two-dimensional centered rectangular lattice fulfill the relation $b/a = 1/2$, a given lattice point of this lattice has 2 NNs, 4 NNNs, 4 TNNs, and 4 fourth nearest neighbors.

Exercise 2.10

Ad (a) & (b) A lattice point in the *fcc* lattice has 12 NNs, 6 NNNs, and 24 TNNs. Figure 2.50 shows all the NNs, 1/3 of the NNNs, and also 1/3 of the TNNs of the lattice point located at the center of the draw. Another 1/3 of the NNNs and 1/3 of the TNNs are located in the empty cubes shown in Fig. 2.50. The rest of the NNNs and TNNs occupy the volume which will be obtained after rotating the solid figure (composed of 6 cubes, see Fig. 2.50) around the vertical *fcc* fourfold rotation axis, which contains the geometric center of the figure, by an angle $2\pi/4$.

Exercise 2.11

Ad (a) The distance to the origin O of the lattice point of coordinates n_1, n_2 is expressed by the formula $d = a\sqrt{n_1^2 + n_2^2}$.

Ad (b) There is a total of 108 neighbors of the lattice point placed at the origin whose distances fulfill the relation $d < 6a$. Those are NNs, NNNs, and so on up to seventeenth nearest neighbors. The information about the neighbors up to order seven is given in Exercise 2.12.

Ad (c) In the case of 12 lattice points at the same distance d from the origin, 3 of them belong to the region shown in Fig. 2.51 and the rest can be obtained by rotations. For instance, in the case of thirteenth nearest neighbors of the lattice point placed at the origin, the coordinates of the 3 points are

$$n_1, n_2 = 5, 0; n_1', n_2' = 4, 3; n_1'', n_2'' = 3, 4. \qquad \text{(A.3)}$$

Exercise 2.12

Ad (a) As an example let us consider the points indexed by $n_1, n_2, \pm3$. Among these points there are no NNs up to order seventh, but there are 2 eighth nearest neighbors at a distance of $a\sqrt{9}$ and 4 ninth nearest neighbors at a distance of $a\sqrt{10}$ to the origin.

Ad (b) There are no lattice points at a distance of $a\sqrt{7}$ from the origin, since the relation $n_1^2 + n_2^2 + n_3^2 = 7$ is not fulfilled by any three integer numbers n_1, n_2, n_3.

Exercise 2.13 By using two fourfold rotation axes, we can obtain the 11 NNNs of the central point in Fig. 2.24, starting with one NNN and doing, first, 2 consecutive rotations by an angle of $2\pi/4$ about a fourfold rotation axis selected in the way that the 2 new NNNs belong to the same face (of the large cube) as the starting point. Next, by doing 3 more similar rotations about a properly selected one of the two remaining fourfold rotation axes, we can obtain the remaining 9 NNNs from the 3 obtained before.

Exercise 2.14 We can see in Fig. 2.52 that the body diagonal of the cube coincides with the hexagonal prism height of length c and the cube face diagonal has the length of the hexagon side a. Let us denote a_c as the cube edge length. Then

$$\begin{cases} c = \sqrt{3}a_c \\ a = \sqrt{2}a_c \end{cases}. \tag{A.4}$$

Finally, from the above we obtain $c/a = \sqrt{6}/2$.

Exercise 2.15 To help yourself

Ad (a) See Fig. 2.52.
Ad (b) See Fig. 2.24.

Exercise 2.16

Ad (b) In your draw you have 9 of the 12 NNs of the lattice point in consideration. This can be verified by comparing the rhombohedral P and triple hexagonal R cell parameters a_r and a_h given in Table 2.7.

Ad (c) Notice that if vector $\vec{R}$ defined by Eq. 2.1 gives the position of a NN then vector $-\vec{R}$ gives the position of another NN of the same lattice point.

Exercise 2.17 For instance, 2 of the NNs of the lattice point placed at the origin **O** of the three unit cells drawn in Fig. 2.41 have positions $\vec{c}_h$ and $-\vec{c}_h$. To find other NNs, we should compare the rhomboheral P and triple hexagonal R cell parameters a_r and c_h given in Table 2.6. See also the response to Exercise 2.16c.

Exercise 2.18 The NNNs, TNNs, and fourth nearest neighbors of the lattice point that is placed at the origin **O** may be divided in 4 groups: those inside/outside the hexagonal prism and those defining the front/back base of the prism.

Ad (a) Half of the NNNs is included in the draw. All of them are inside of the hexagonal prism.

Ad (b) Three quarters of the TNNs is included in the draw. Six of them define the hexagonal prism front base and 3 of them are outside of the prism. Finally, 1/4 of the fourth nearest neighbors is included in the draw. All of them define the back base of the prism.

Exercise 2.19

Ad (b) The volume of the hexagonal prism is 9 times the volume of the rhombohedron. To calculate the volume of the rhombohedron use the formula included in Fig. 2.1. The components of the basis vectors $\vec{a}_1$, $\vec{a}_2$, and $\vec{a}_3$ can be found using Fig. 2.53.

Exercise 2.20

Ad (b) To obtain the c/a ratio given by Eq. 2.12, start from the second equation of Eqs. 2.11.

Exercise 2.21 As an example, we show that $(c/a)_{sc} = \sqrt{6}/2$, namely

$$(c/a)_{sc} = \sqrt{\frac{9}{4\sin^2(90°/2)} - 3} = \sqrt{\frac{9}{4\left(\sqrt{2}/2\right)^2} - 3} = \sqrt{\frac{3}{2}} = \frac{\sqrt{6}}{2}.$$

$$(A.5)$$

Exercise 2.22

Ad (b) The distance between NN atoms in α-Hg at 227 K is 3.005 Å.

Exercise 2.23 To calculate the angles between the axes that define the primitive rhombohedral unit cell of the *bcc* lattice follow the steps described in Section 2.10 for the *fcc* lattice.

Exercise 2.24 See Fig. 2.34.

Exercise 2.25

Ad (a) The rhombohedral unit cell described in this exercise has 4 lattice points. To be sure that we have taken into account just the lattice points that belong to the rhombohedron, it is useful to calculate

(1) the ratio between the volumes of the rhombohedron and the large cube from Fig. 2.34,
(2) the number of *bcc* lattice points that belong to the large cube.

Multiplying the volume ratio by the number obtained in point (2), we obtain the result.

Exercise 2.26 Your drawing should be similar to that in Fig. 2.55, but remember that in this case we are in presence of the *sc*, not the *fcc*, lattice.

Ad (a) In the *sc* lattice, the smallest rhombohedral unit cell defined by basis vectors with an angle of 109°28' between them is a rhombohedral *F* unit cell.

Ad (b) Your answer should be 4.

Ad (c) To calculate the volume of the rhombohedral unit cell use the formula included in Fig. 2.1. The expressions for the basis vectors $\vec{a}_1$, $\vec{a}_2$, and $\vec{a}_3$ which define the rhombohedral cell can be proposed in a similar way as it was done in Fig. 2.26.

Exercise 2.27

Ad (a) Your answer should be 16. To be sure that you have taken into account all the lattice points that belong to the rhombohedral unit cell, follow the steps indicated in the response for Exercise 2.25.

Ad (b) Two vertices of the cube coincide with the two vertices of the rhombohedral unit cell which define its shorter body diagonal and the other six are placed at the geometric centers of the rhombohedron faces.

Exercise 2.28 For drawing, make use of Fig. 2.48. In this figure, the additional points at the edges and at the center of the large cube, that do not belong to the *fcc* lattice, can be used as a guide to the eye for constructing the hexagonal prism whose top or bottom base is of the shape shown in the figure. To form the prism, two replicas of the hexagon from the figure should be translated along the cube diagonal in such a way that their centers coincide with two of the large cube vertices. Next, to draw the rhombohedral *P* cell, make use of Fig. 2.31.

Exercise 2.29 The basis vectors of a *C*-centered orthorhombic unit cell of a hexagonal lattice are $\vec{a}_{ortho} = 2\vec{a}_h + \vec{b}_h$, $\vec{b}_{ortho} = \vec{b}_h$, and

$\vec{c}_{ortho} = \vec{c}_h$, where $\vec{a}_h$, $\vec{b}_h$, and $\vec{c}_h$ define a hexagonal unit cell of the hexagonal lattice.

Exercise 2.30

Ad (a) See the response for Exercise 2.29.
Ad (b) Such orthorhombic unit cell has 6 centering points.

Exercise 2.31

Ad (a) Only the NNs participate in the construction of the Wigner-Seitz cell of a two-dimensional hexagonal lattice.
Ad (b) The NNs and NNNs participate in the construction of the Wigner–Seitz cell of the two-dimensional lattice defined in Exercise 2.9a.

Exercise 2.32 See the response for Exercise 2.31a.

Exercise 2.33 Figure A.27 shows a vertical plane defined by two body diagonals of the small cube of edge a from Fig. 2.46. In this figure, we have shown only the cross section of the small cube and the cross section of the octahedron. The parameter x, defined in the draw, represents half of the edge length of the larger cube and the

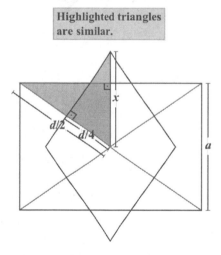

Figure A.27 Vertical plane defined by body diagonals of the small cube from Fig. 2.46. The cross section of the octahedron is also shown.

parameter $d = \sqrt{3}a$ corresponds to the length of the small cube diagonal. The two highlighted triangles are similar, so

$$\frac{x}{d/2} = \frac{d/4}{a/2}. \qquad (A.6)$$

From Eq. A.6, we obtain the parameter x as equal to $(3/4)\,a$ and the edge length $2x$ of the larger cube is equal to $(3/2)\,a$.

A.3 Solutions to Exercises of Chapter 3

Exercise 3.1

Ad (b) The filling factor for the diamond structure is 0.34.

Ad (c) The filling factor of the diamond structure is the smallest one if compared with the filling factors of the *fcc*, *bcc*, and *sc* structures. Its value represents half of the filling factor for the *bcc* structure and only 46% of the filling factor for the *fcc* structure, which has the highest possible value among all the filling factors.

Exercise 3.2 The twofold rotation axes of a regular tetrahedron are sub-elements of its rotoinversion axes $\bar{4}$.

Exercise 3.3

Ad (a) & (b) The set of atoms shown in Fig. 3.35a has $\bar{3}$ as the highest-order symmetry axis, while the highest-order symmetry axis of the set of atoms shown in Fig. 3.35b is $\bar{6}$.

Exercise 3.4 The centers of symmetry lie on layers **A**.

Exercise 3.5 Draw a dashed line, which indicates how the front atom from the bottom base of the hexagonal prism changes its position when passing from layer **A** to layer **B** after right-handed screw rotation of 180° around the 2_1 axis. Next, as a result of the same screw rotation the back atom from layer **B** takes place of the front atom from the top base of the hexagonal prism.

Exercise 3.6

Ad (a) The origin of the hexagonal P unit cell is lying at the center of symmetry located on the $6_3/m$ symmetry axis of the infinite *dhcp* structure. In the site symmetry symbol of the origin is included the rotoinversion axis $\bar{3}$. This is one of the two alternative origin choices in the case of the *dhcp* structure (neither of them is considered standard).

Ad (b) The coordinate triplets of the four atoms within the hexagonal P unit cell drawn in (a) are: $0, 0, 0$; $\frac{2}{3}, \frac{1}{3}, \frac{1}{4}$; $0, 0, \frac{1}{2}$; $\frac{1}{3}, \frac{2}{3}, \frac{3}{4}$.

Exercise 3.7

Ad (c) The coordinate triplets of the 6 atoms, within the hexagonal P unit cell with the origin at the center of an atom on layer **A**, are $0, 0, 0$; $\frac{2}{3}, \frac{1}{3}, \frac{1}{6}$; $\frac{1}{3}, \frac{2}{3}, \frac{1}{3}$; $\frac{2}{3}, \frac{1}{3}, \frac{1}{2}$; $0, 0, \frac{2}{3}$; $\frac{1}{3}, \frac{2}{3}, \frac{5}{6}$.

Ad (d) The c/a ratio is 3×1.639, which is close to the ideal value of 3×1.633.

Ad (e) The 6 NNs of an atom, which are placed in the adjacent layers, are at a distance of 2.917 Å to this atom. It means, less than 0.3% farther than the 6 NNs that are within the same layer.

Exercise 3.8

Ad (b) The volume of the triple hexagonal R unit cell is three times the volume of the rhombohedral P unit cell. As a reminder, the ratio between the volumes is equal to the ratio between the numbers of atoms belonging to them.

Ad (c) The c_h/a_h ratio is $(9/2) \times 1.605$, which is 2% smaller than the ideal value of $(9/2) \times 1.633$.

Exercise 3.9 In the case of the α-Sm structure the coordinate triplets of the nine atoms within a triple hexagonal R unit cell in obverse setting are: $0, 0, 0$; $\frac{2}{3}, \frac{1}{3}, \frac{1}{9}$; $0, 0, \frac{2}{9}$; $\frac{2}{3}, \frac{1}{3}, \frac{1}{3}$; $\frac{1}{3}, \frac{2}{3}, \frac{4}{9}$; $\frac{2}{3}, \frac{1}{3}, \frac{5}{9}$; $\frac{1}{3}, \frac{2}{3}, \frac{2}{3}$; $0, 0, \frac{7}{9}$; $\frac{1}{3}, \frac{2}{3}, \frac{8}{9}$.

Exercise 3.10 A set of atoms, within the crystal structure of α-As, that has a rotoinversion axis $\bar{3}$ is shown in Fig. A.28.

Exercise 3.11 At normal conditions, ytterbium in the α phase has 6 NNs at a distance of 3.8799 Å within the same layer, and another 6

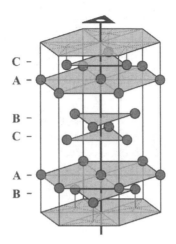

C –
A –

B –
C –

A –
B –

Figure A.28 Hexagonal prism for the α-As crystal structure. The set of atoms shown in the draw has a rotoinversion axis $\bar{3}$. The inversion point coincides with the center of mass.

neighbors at a distance of 3.9004 Å in adjacent layers. All the 12 NNs are on average on a distance of 3.8901 Å from a given atom. On the other hand, β-Yb at normal conditions has its 12 NNs at a distance of 3.8783 Å. The average NN interatomic distance in α-Yb is therefore larger by only 0.3% than the NN interatomic distance in β-Yb.

Exercise 3.12 At normal conditions, an atom of cerium in the β phase has its 6 NNs in adjacent layers at a distance of 3.647 Å, and another 6 neighbors in the same layer at a distance of 3.681 Å, which is only slightly larger. The average NN interatomic distance is 3.664 Å. On the other hand, an atom in the γ-Ce phase (at normal conditions) has its 12 NNs at a distance of 3.6494 Å. The average NN interatomic distance in β-Ce is therefore only about 0.4% larger than the NN interatomic distance in γ-Ce.

Exercise 3.13 The NN interatomic distance in α-Fe at ambient conditions is 2.4825 Å, while the interatomic distance in δ-Fe at 1712 K and normal pressure is 2.5414 Å. Therefore, at normal pressure, the interatomic distance in Fe increases by 2.4% when the temperature changes from room temperature to 1712 K.

Exercise 3.14 The NN interatomic distances for five elements (Ca, Ce, Fe, Li, and Na) in the *bcc* and *fcc* or *hcp* structures are compared in Table A.2.

Table A.2 Nearest neighbor distances, d_{NN}, for five elements in the *bcc* and *fcc* or *hcp* structures. The interatomic distances are given in Angstroms

Element	d_{NN} (temperature)		$\frac{d_{NN} - d_{NN}^{bcc}}{d_{NN}^{bcc}}$
	bcc structure	*fcc* or *hcp* structure	
Ca	3.793 (773 K)	3.9516 (r.t.)	4%
Ce	3.568 (1030 K)	3.6494 (r.t.)	2%
Fe	2.5414 (1712 K)	2.567 (1373 K)	1%
Li	3.0391 (r.t.)	$d_{NN}^{I} = 3.111, d_{NN}^{II} = 3.116$ (78 K)	2.4%, 2.5%
Na	3.716 (r.t.)	$d_{NN}^{I} = 3.767, d_{NN}^{II} = 3.768$ (5 K)	1.4%, 1.4%

Exercise 3.15

Ad (a) The distances from the central atom to its NNs, NNNs, TNNs, fourth nearest neighbors, and fifth nearest neighbors are a_h, $\sqrt{2}a_h$, $\sqrt{3}a_h$, $\sqrt{4}a_h$, and $\sqrt{5}a_h$, respectively.

Ad (b) The number of NNs, NNNs, TNNs, and fourth nearest neighbors is 12, 6, 24, and 12, respectively.

A.4 Solutions to Exercises of Chapter 4

Exercise 4.1 The zinc blende (β-ZnS) structure has 4 cations and 4 anions in the cubic F unit cell. In the cell with a Zn cation at the origin the coordinate triplets for the Zn cations are $0, 0, 0$; $\frac{1}{2}, \frac{1}{2}, 0$; $\frac{1}{2}, 0, \frac{1}{2}$; $0, \frac{1}{2}, \frac{1}{2}$ and for the S anions are $\frac{1}{4}, \frac{1}{4}, \frac{1}{4}$; $\frac{3}{4}, \frac{3}{4}, \frac{1}{4}$; $\frac{3}{4}, \frac{1}{4}, \frac{3}{4}$; $\frac{1}{4}, \frac{3}{4}, \frac{3}{4}$.

Exercise 4.2 The CaF$_2$ structure has 4 cations and 8 anions in the cubic F unit cell. In the cell with a Ca^{2+} cation at the origin the coordinate triplets for the Ca^{2+} cations are $0, 0, 0$; $\frac{1}{2}, \frac{1}{2}, 0$; $\frac{1}{2}, 0, \frac{1}{2}$; $0, \frac{1}{2}, \frac{1}{2}$ and for the F$^-$ anions are $\frac{1}{4}, \frac{1}{4}, \frac{1}{4}$; $\frac{3}{4}, \frac{1}{4}, \frac{1}{4}$; $\frac{1}{4}, \frac{3}{4}, \frac{1}{4}$; $\frac{3}{4}, \frac{3}{4}, \frac{1}{4}$; $\frac{1}{4}, \frac{1}{4}, \frac{3}{4}$; $\frac{3}{4}, \frac{1}{4}, \frac{3}{4}$; $\frac{1}{4}, \frac{3}{4}, \frac{3}{4}$; $\frac{3}{4}, \frac{3}{4}, \frac{3}{4}$.

Table A.3 Next nearest neighbor distances, d_{NNN}, of an ion for five compounds in the zinc blende and wurtzite structures. The interatomic distances are given in Angstroms

Compound	d_{NNN}			$\left|\frac{d_{NNN}^{WZ} - d_{NNN}^{ZB}}{d_{NNN}^{ZB}}\right|$	
	Zinc blende (ZB)	Wurtzite (WZ)			
		In-plane	Out-of-plane	In-plane	Out-of-plane
MnSe	4.173	4.12	4.12	1.3%	1.3%
MnTe	4.482	4.48	4.48	0%	0%
ZnSe	4.0076	4.003	4.004	0.1%	0.1%
CdSe	4.297	4.2999	4.2955	0.1%	0%
GaN	3.190	3.1878	3.1794	0.1%	0.3%

Exercise 4.3 The results are given in Table A.3. In all five compounds, the c/a ratio of the wurtzite structure is close to the ideal value of 1.633, so each ion has 12 NNNs (6 in-plane and 6 out-of-plane).

Exercise 4.4

Ad (a) A hexagonal prism for the β-ZnS (zinc blende) structure is shown in Fig. A.29.

Ad (b) To the prism from (a) belong 9 ions of each type, while to the hexagonal prism from Fig. 4.15 (wurtzite structure) belong 6 ions of each type.

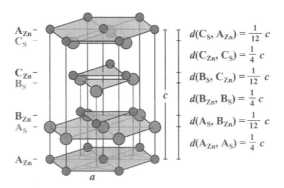

$$d(C_S, A_{Zn}) = \frac{1}{12}c$$
$$d(C_{Zn}, C_S) = \frac{1}{4}c$$
$$d(B_S, C_{Zn}) = \frac{1}{12}c$$
$$d(B_{Zn}, B_S) = \frac{1}{4}c$$
$$d(A_S, B_{Zn}) = \frac{1}{12}c$$
$$d(A_{Zn}, A_S) = \frac{1}{4}c$$

Figure A.29 Hexagonal prism for β-ZnS, which crystallizes in the zinc blende structure. Two-dimensional *hcp* layers A_{Zn}, B_{Zn}, C_{Zn}, A_S, B_S, and C_S are shown. In addition, the draw shows the distances between the consecutive layers.

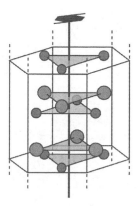

Figure A.30   A sixfold screw axis 6_3 in an infinite wurtzite structure.

Exercise 4.5   A sixfold screw axis 6_3 of an infinite wurtzite structure is shown in Fig. A.30.

Exercise 4.6   The set of ions shown in Fig. 4.18a to the left has $\bar{6}$ rotoinversion axis as the highest-order symmetry axis and the rotoinversion axis $\bar{3}$ is the highest-order symmetry axis for the set of ions shown in Fig. 4.18a to the right.

Exercise 4.7   A sixfold screw axis with a center of symmetry, $6_3/m$, in the infinite volume of the NiAs structure is shown in Fig. A.31.

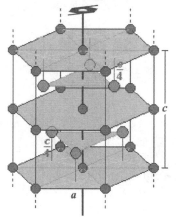

Figure A.31   A sixfold screw axis with a center of symmetry, $6_3/m$, in the infinite volume of the NiAs structure.

Table A.4 The interatomic distances, given in Angstroms, for NiAs type structures with wide range of lattice constant ratios c/a. The parameters d_{cc}^{I} and d_{cc}^{II} denote the two closest cation-cation distances and d_{ac} denotes the distance of a cation to its nearest anions. The number of anion NNs of a cation is labeled as n_{ac}, while n_{cc}^{I} and n_{cc}^{II} denote the numbers of cations at distances d_{cc}^{I} and d_{cc}^{II} from a cation, respectively

Compound (c/a)	d_{cc}^{I}, d_{cc}^{II}	$\left\|\frac{d_{cc}^{II}-d_{cc}^{I}}{d_{cc}^{I}}\right\|$	d_{ac}	$\left\|\frac{d_{cc}^{I}-d_{ac}}{d_{ac}}\right\|$	Number of NNs of a cation	Number of NNNs of a cation
VSb (1.28)	2.724, 4.27	57%	2.82	3%	$n_{ac} + n_{cc}^{I} = 8$	$n_{cc}^{II} = 6$
VSe (1.63)	2.98, 3.66	23%	2.58	16%	$n_{ac} = 6$	$n_{cc}^{I} = 2$
VS (1.75)	2.91, 3.33	14%	2.41	21%	$n_{ac} = 6$	$n_{cc}^{I} = 2$
TiS (1.93)	3.190, 3.299	3%	2.484	28%	$n_{ac} = 6$	$n_{cc}^{I} + n_{cc}^{II} = 8$

Exercise 4.8

Ad (a) The results are given in Table A.4. When calculating the number of NNNs of a cation, we assume that whenever d_{cc}^{I} and d_{cc}^{II} differ by less than 10%, then the ions at both distances are NNNs. On the other hand, if the shortest distance d_{cc}^{I} differs from the distance of a cation to its nearest anions, d_{ac}, by less than 10%, then the cations at such a distance join the group of NNs of the cation in consideration.

Ad (b) To determine the number of the NNNs of a given cation use the criterions suggested in the response for (a).

Exercise 4.9

Ad (a) & (b) The results are given in Table A.5.

Exercise 4.10

Ad (a) The distances between NNs in α-ZrP and β-ZrP are $d_{ac}^{\alpha} = 2.632$ Å and $d_{ac}^{\beta} = 2.643$ Å, respectively. They differ in value by 0.4%.

Ad (b) The results are given in Table A.6.

Ad (c) The results are given in Table A.7.

Table A.5 The interatomic distances, given in Angstroms, for NiAs type structures with lattice constant ratios c/a much smaller than 1.633. The parameter d_{ac} denotes the distance of a cation to its 6 nearest anions and d_{cc} denotes the distance of a cation to its 2 nearest cations, while $d_{XX}^{element}$ denotes the interatomic distance for the element. The subscript X in $d_{XX}^{element}$ represents Cu, Pd, or Ir

Compound	c/a	d_{ac}	d_{cc}	$\left\vert\frac{d_{cc}-d_{ac}}{d_{ac}}\right\vert$	$d_{XX}^{element}$	$\left\vert\frac{d_{cc}-d_{XX}^{element}}{d_{XX}^{element}}\right\vert$
CuSb	1.34	2.586	2.597	0.4%	2.556	1.6%
PdSb	1.37	2.738	2.797	2.2%	2.751	1.7%
IrSb	1.39	2.680	2.761	3.0%	2.715	1.7%
IrTe	1.37	2.643	2.693	1.9%	2.715	0.8%

Table A.6 The distances d_{aa} between an anion and its NNNs (12 anions) in the α-ZrP and β-ZrP structures. The interatomic distances are given in Angstroms

	α-ZrP	β-ZrP	
		12 in total	
Number of NNNs of an anion	12	6	6
d_{aa}	3.722	3.684	3.791
$\left\vert\frac{d_{aa}-d_{aa}^{\alpha}}{d_{aa}^{\alpha}}\right\vert$	0%	1%	1.8%

Table A.7 The distance d_{cc} between a cation and its NNNs (12 cations) in the α-ZrP structure and the distances between a cation and 10 closest cations to it in the β-ZrP structure. The interatomic distances are given in Angstroms

	α-ZrP	β-ZrP		
		10 in total		
Number of NNNs of a cation	12	1	6	3
d_{cc}	3.722	3.139	3.684	3.791
$\left\vert\frac{d_{cc}-d_{cc}^{\alpha}}{d_{cc}^{\alpha}}\right\vert$	0%	15.7%	1%	1.8%

Exercise 4.11

Ad (a) The results are given in Table A.8. In all three structures, each ion has 6 NNs.

Table A.8 Ion-ion NN distances d_{ac} in the δ-NbN, δ'-NbN, and ε-NbN structures. The interatomic distances are given in Angstroms

	δ-NbN	δ'-NbN	ε-NbN
d_{ac}	2.197	2.205	2.209
$\left\|\frac{d_{ac}-d_{ac}^{\delta}}{d_{ac}^{\delta}}\right\|$	0%	0.4%	0.5%

Ad (b) The results are given in Table A.9.

Table A.9 The distances d_{aa} between an anion and its NNNs in the δ-NbN, δ'-NbN, and ε-NbN structures. The interatomic distances are given in Angstroms

	δ-NbN	δ'-NbN		ε-NbN	
		8 in total		12 in total	
Number of NNNs of an anion	12	6	2	6	6
d_{aa}	3.107	2.968	2.775	2.9513	3.288
$\left\|\frac{d_{aa}-d_{aa}^{\delta}}{d_{aa}^{\delta}}\right\|$	0%	4.5%	10.7%	5%	5.8%

Ad (c) The results are given in Table A.10. To calculate the NNN distances in δ'-NbN draw a hexagonal prism with the Nb cations in its vertices to visualize better the location of the NNNs of a cation in the anti-NiAs structure.

Table A.10 The distances d_{cc} between a cation and its NNNs in the δ-NbN, δ'-NbN, and ε-NbN structures. The interatomic distances are given in Angstroms

	δ-NbN	δ'-NbN		ε-NbN		
		12 in total		10 in total		
Number of NNNs of a cation	12	6	6	1	6	3
d_{cc}	3.107	2.968	3.261	2.812	2.9513	3.288
$\left\|\frac{d_{cc}-d_{cc}^{\delta}}{d_{cc}^{\delta}}\right\|$	0%	4.5%	5%	9.5%	5%	5.8%

Table A.11 Filling factors and ionic radius ratios for alkali halides and for the *sc* structure. Note that for the increasing radius ratio (r_+/r_- or r_-/r_+) the filling factor is decreasing

Compound	r_+/r_- or r_-/r_+	Filling factor
LiCl	0.42	0.79
NaCl	0.56	0.65
RbCl	0.84	0.56
KF	0.96	0.55
sc structure	1	0.52

Exercise 4.12

Ad (a) & (b) The results are given in Table A.11. When $r_+/r_- = 1$, we have the case of the *sc* structure whose filling factor is 0.52.

Exercise 4.14 Such unit cell is defined by cations.

A.5 Solutions to Exercises of Chapter 5

Exercise 5.1

Ad (a) If the translation vector $\vec{b}$ of the reciprocal lattice is expressed as follows

$$\vec{b} = \alpha \hat{b}_1 + \beta \hat{b}_2 \tag{A.7}$$

we have

$$\begin{aligned}
\vec{b} \cdot \vec{a}_1 &= \alpha \hat{b}_1 \cdot \vec{a}_1 + \beta \hat{b}_2 \cdot \vec{a}_1 = \alpha \hat{b}_1 \cdot \vec{a}_1 \\
\vec{b} \cdot \vec{a}_2 &= \alpha \hat{b}_1 \cdot \vec{a}_2 + \beta \hat{b}_2 \cdot \vec{a}_2 = \beta \hat{b}_2 \cdot \vec{a}_2
\end{aligned} \tag{A.8}$$

as $\hat{b}_2 \perp \vec{a}_1$ and $\hat{b}_1 \perp \vec{a}_2$. Taking into account Eqs. 5.40, we have

$$\begin{aligned}
\alpha \hat{b}_1 \cdot \vec{a}_1 &= 2\pi g_1 \Rightarrow \alpha = \frac{2\pi}{\hat{b}_1 \cdot \vec{a}_1} g_1 = \frac{2\pi}{a_1 \sin \sphericalangle (\vec{a}_1, \vec{a}_2)} g_1 \\
&= \frac{2\pi a_2}{a_1 a_2 \sin (\vec{a}_1, \vec{a}_2)} g_1 = \frac{2\pi a_2}{\Omega_0} g_1,
\end{aligned} \tag{A.9}$$

where Ω_0 is the volume of the primitive unit cell. In a similar way can be obtained the coefficient β

$$\beta = \frac{2\pi a_1}{\Omega_0} g_2, \tag{A.10}$$

so

$$\vec{b} = \alpha \hat{b}_1 + \beta \hat{b}_2 = g_1 \frac{2\pi a_2}{\Omega_0} \hat{b}_1 + g_2 \frac{2\pi a_1}{\Omega_0} \hat{b}_2 \tag{A.11}$$

and we have the following expressions for the primitive translation vectors $\vec{b}_1$ and $\vec{b}_2$

$$\begin{cases} \vec{b}_1 = \dfrac{2\pi a_2}{\Omega_0} \hat{b}_1, \ \hat{b}_1 \perp \vec{a}_2 \\ \vec{b}_2 = \dfrac{2\pi a_1}{\Omega_0} \hat{b}_2, \ \hat{b}_2 \perp \vec{a}_1 \end{cases}. \tag{A.12}$$

The translation vector $\vec{b}$ of the reciprocal lattice is expressed by $\vec{b}_1$ and $\vec{b}_2$ as follows

$$\vec{b} = g_1 \vec{b}_1 + g_2 \vec{b}_2. \tag{A.13}$$

Exercise 5.2 From Eqs. A.12 and Eq. 1.2, we have

$$\vec{a}_1 \cdot \vec{b}_1 = \frac{2\pi a_2}{\Omega_0} \vec{a}_1 \cdot \hat{b}_1 = \frac{2\pi a_2 a_1 \cos \sphericalangle \left(\vec{a}_1, \hat{b}_1 \right)}{\Omega_0}$$

$$= \frac{2\pi a_2 a_1 \sin \sphericalangle (\vec{a}_1, \vec{a}_2)}{\Omega_0} = 2\pi, \tag{A.14}$$

since $\sphericalangle \left(\vec{a}_1, \hat{b}_1 \right) = 90° - \sphericalangle (\vec{a}_1, \vec{a}_2)$ (see Fig. 5.8) and $a_2 a_1 \sin \sphericalangle (\vec{a}_1, \vec{a}_2) = \Omega_0$. As $\vec{b}_1$ is orthogonal to $\vec{a}_2$, we have

$$\vec{a}_2 \cdot \vec{b}_1 = 0. \tag{A.15}$$

Similarly, we can prove the rest of cases.

Exercise 5.3

Ad (a) The basis vectors which define a rhombohedral primitive unit cell of the *bcc* lattice are shown in Fig. 2.26 in Section 2.9.

The primitive translation vectors of the lattice which is reciprocal to the *bcc* lattice are as follows:

$$\vec{b}_1 = 2\pi \frac{\vec{a}_2 \times \vec{a}_3}{\Omega_0} = \frac{2\pi}{\frac{1}{2}a^3} \begin{vmatrix} \hat{x} & \hat{y} & \hat{z} \\ \frac{1}{2}a & \frac{1}{2}a & -\frac{1}{2}a \\ -\frac{1}{2}a & \frac{1}{2}a & \frac{1}{2}a \end{vmatrix} = \frac{2\pi}{a}\hat{x} + \frac{2\pi}{a}\hat{z}$$

$$\vec{b}_2 = 2\pi \frac{\vec{a}_3 \times \vec{a}_1}{\Omega_0} = \frac{2\pi}{\frac{1}{2}a^3} \begin{vmatrix} \hat{x} & \hat{y} & \hat{z} \\ -\frac{1}{2}a & \frac{1}{2}a & \frac{1}{2}a \\ \frac{1}{2}a & -\frac{1}{2}a & \frac{1}{2}a \end{vmatrix} = \frac{2\pi}{a}\hat{x} + \frac{2\pi}{a}\hat{y}.$$

$$\vec{b}_3 = 2\pi \frac{\vec{a}_1 \times \vec{a}_2}{\Omega_0} = \frac{2\pi}{\frac{1}{2}a^3} \begin{vmatrix} \hat{x} & \hat{y} & \hat{z} \\ \frac{1}{2}a & -\frac{1}{2}a & \frac{1}{2}a \\ \frac{1}{2}a & \frac{1}{2}a & -\frac{1}{2}a \end{vmatrix} = \frac{2\pi}{a}\hat{y} + \frac{2\pi}{a}\hat{z}$$

$$(A.16)$$

We can see that $\vec{b}_1$, $\vec{b}_2$, $\vec{b}_3$ represent indeed the primitive translation vectors of the *fcc* lattice with lattice constant $4\pi/a$.

Exercise 5.4 When we substitute the expressions for $\vec{b}_2$ and $\vec{b}_3$ given by Eq. 5.29 into $2\pi \left(\vec{b}_2 \times \vec{b}_3 \right)/V$, we obtain

$$2\pi \frac{\left(\vec{b}_2 \times \vec{b}_3 \right)}{V} = \frac{(2\pi)^3}{V} \left[\frac{(\vec{a}_3 \times \vec{a}_1)}{\Omega_0} \times \frac{(\vec{a}_1 \times \vec{a}_2)}{\Omega_0} \right]$$

$$= \frac{1}{\Omega_0} [(\vec{a}_3 \times \vec{a}_1) \times (\vec{a}_1 \times \vec{a}_2)], \qquad (A.17)$$

since the volume of the unit cell of the reciprocal lattice is expressed by the volume of the unit cell of the direct lattice as $V = (2\pi)^3/\Omega_0$. By making use of the vector identity given in the exercise content, we have ($\vec{A} = \vec{a}_3 \times \vec{a}_1, \vec{B} = \vec{a}_1$, and $\vec{C} = \vec{a}_2$)

$$(\vec{a}_3 \times \vec{a}_1) \times (\vec{a}_1 \times \vec{a}_2) = \vec{a}_1 [(\vec{a}_3 \times \vec{a}_1) \cdot \vec{a}_2] - \vec{a}_2 [(\vec{a}_3 \times \vec{a}_1) \cdot \vec{a}_1]$$

$$= \vec{a}_1 [(\vec{a}_3 \times \vec{a}_1) \cdot \vec{a}_2] = \vec{a}_1 [(\vec{a}_1 \times \vec{a}_2) \cdot \vec{a}_3]$$

$$= \Omega_0 \vec{a}_1. \qquad (A.18)$$

Finally, we obtain

$$2\pi \frac{\left(\vec{b}_2 \times \vec{b}_3 \right)}{V} = \vec{a}_1 \qquad (A.19)$$

and in a similar way it can be shown that

$$2\pi \frac{\left(\vec{b}_3 \times \vec{b}_1 \right)}{V} = \vec{a}_2 \text{ and } 2\pi \frac{\left(\vec{b}_1 \times \vec{b}_2 \right)}{V} = \vec{a}_3. \qquad (A.20)$$

A.6 Solutions to Exercises of Chapter 6

Exercise 6.1

Ad (a) See Figs. 2.8 and 2.48 (Section 2.4 and Exercise 2.5, respectively).

Ad (b) There are two (221) *sc* lattice planes that contain at least two lattice points from the (010) front face of the large cube. None of them is the closest one to the origin.

Exercise 6.2 In Fig. 6.12, we can see that the primitive translation vectors of the lattice which is reciprocal to the *sc* lattice (of lattice constant a) are $\vec{b}_1 = \frac{2\pi}{a}\hat{x}$, $\vec{b}_2 = \frac{2\pi}{a}\hat{y}$, and $\vec{b}_3 = \frac{2\pi}{a}\hat{z}$. Since vector $\vec{b}_{221}$ and its modulus b_{221} are expressed as follows:

$$\vec{b}_{221} = 2\vec{b}_1 + 2\vec{b}_2 + 1\vec{b}_3$$

$$b_{221} = \frac{2\pi}{a}\sqrt{4+4+1} = 3\frac{2\pi}{a}, \tag{A.21}$$

thus from Eq. 6.11 we obtain

$$d_{221} = \frac{2\pi}{b_{221}} = \frac{2\pi}{6\pi/a} = \frac{a}{3}. \tag{A.22}$$

Exercise 6.3

Ad (a) Draw such lattice planes inside the cubic unit cells (of edge a) of the *sc* and *fcc* lattices, P and F, respectively.

Ad (b) The primitive translation vectors of the lattice which is reciprocal to the *fcc* lattice defined by primitive translation vectors $\vec{a}_1 = \frac{1}{2}a\hat{x} + \frac{1}{2}a\hat{z}$, $\vec{a}_2 = \frac{1}{2}a\hat{x} + \frac{1}{2}a\hat{y}$, and $\vec{a}_3 = \frac{1}{2}a\hat{y} + \frac{1}{2}a\hat{z}$ are

$$\vec{b}_1 = \frac{2\pi}{a}\hat{x} - \frac{2\pi}{a}\hat{y} + \frac{2\pi}{a}\hat{z}$$

$$\vec{b}_2 = \frac{2\pi}{a}\hat{x} + \frac{2\pi}{a}\hat{y} - \frac{2\pi}{a}\hat{z}$$

$$\vec{b}_3 = -\frac{2\pi}{a}\hat{x} + \frac{2\pi}{a}\hat{y} + \frac{2\pi}{a}\hat{z} \tag{A.23}$$

(see Figs. 5.6 and 5.7 in Sections 5.3.3 and 5.3.4, respectively). The vector $\vec{b}_{111}$ is expressed as follows:

$$\vec{b}_{111} = \vec{b}_1 + \vec{b}_2 + \vec{b}_3 = \frac{2\pi}{a}\hat{x} + \frac{2\pi}{a}\hat{y} + \frac{2\pi}{a}\hat{z}. \tag{A.24}$$

It means, in the same way as $\vec{b}_{111}$ is shown in Fig. 6.13 for the *sc* lattice, in both cases

$$d_{111} = \frac{2\pi}{b_{111}} = \frac{\sqrt{3}}{3}a. \qquad (A.25)$$

Ad (c) If we choose as the basis vectors which define a primitive rhombohedral unit cell of the *bcc* lattice, those shown in Fig. 2.26 (Section 2.9), then the primitive translation vectors of the *fcc* lattice which is reciprocal to the *bcc* lattice are as follows:

$$\vec{b}_1 = \frac{2\pi}{a}\hat{x} + \frac{2\pi}{a}\hat{z}$$
$$\vec{b}_2 = \frac{2\pi}{a}\hat{x} + \frac{2\pi}{a}\hat{y}$$
$$\vec{b}_3 = \frac{2\pi}{a}\hat{y} + \frac{2\pi}{a}\hat{z} \qquad (A.26)$$

(see the solution of Exercise 5.3). Use these translation vectors to calculate the distance between the consecutive planes which are orthogonal to a body diagonal of the cubic unit cell of the *bcc* lattice ($\vec{b}_{111}$ is parallel to the shortest diagonal of the primitive rhombohedral unit cell, and to a diagonal of the cube).

Exercise 6.4

Ad (b) To find the primitive translation vectors $\vec{b}_1$ and $\vec{b}_2$ follow the indications given in Exercise 5.1a and its solution.

A.7 Solutions to Exercises of Chapter 7

Exercise 7.1 The two atoms are placed at $\vec{r}_1 = \vec{0}$ and $\vec{r}_2 = \frac{1}{2}\vec{a}_1 + \frac{1}{2}\vec{a}_2 + \frac{1}{2}\vec{a}_3$, where $\vec{a}_1, \vec{a}_2, \vec{a}_3$ define the cubic P unit cell of the CsCl crystal structure. The square of the modulus of the structure factor $|F(hkl)|^2$ can be obtained in the following way:

$$\begin{aligned}
|F(hkl)|^2 &= F^*(hkl)\,F(hkl) \\
&= \{f_1 + f_2 \exp[-i\pi\,(h+k+l)]\} \\
&\quad \times \{f_1 + f_2 \exp[i\pi\,(h+k+l)]\} \\
&= f_1^2 + f_2^2 + 2f_1 f_2 \cos[\pi\,(h+k+l)], \qquad (A.27)
\end{aligned}$$

where f_1 and f_2 are given by Eq. 7.36. So

$$|F(hkl)|^2 = \begin{cases} (f_1 - f_2)^2, & \text{if } (h+k+l) \text{ is odd} \\ (f_1 + f_2)^2, & \text{if } (h+k+l) \text{ is even} \end{cases} \quad . \qquad \text{(A.28)}$$

In the case of the *bcc* structure $f_1 = f_2 = f$ and the square of the modulus of the structure factor $|F(hkl)|^2$ adopts the following values

$$|F(hkl)|^2 = \begin{cases} 0, & \text{if } (h+k+l) \text{ is odd} \\ 4f^2, & \text{if } (h+k+l) \text{ is even} \end{cases} \quad . \qquad \text{(A.29)}$$

Exercise 7.2

Ad (a) The four atoms within the cubic F unit cell of the *fcc* structure are located at $\vec{r}_1 = \vec{0}$, $\vec{r}_2 = \frac{1}{2}\vec{a}_1 + \frac{1}{2}\vec{a}_2$, $\vec{r}_3 = \frac{1}{2}\vec{a}_1 + \frac{1}{2}\vec{a}_3$, and $\vec{r}_4 = \frac{1}{2}\vec{a}_2 + \frac{1}{2}\vec{a}_3$, where $\vec{a}_1, \vec{a}_2, \vec{a}_3$ define the unit cell. The geometric structure factor S of the *fcc* structure results in

$$\begin{aligned} S(hkl) &= \sum_{m=1}^{4} \exp\left[i2\pi\left(hu_m + kv_m + lw_m\right)\right] \\ &= 1 + \exp\left[i\pi\left(h+k\right)\right] + \exp\left[i\pi\left(h+l\right)\right] \\ &\quad + \exp\left[i\pi\left(k+l\right)\right]. \end{aligned}$$

$$\text{(A.30)}$$

Ad (b)

$$\begin{array}{ll} S(100) = 0 & S(211) = 0 \\ S(110) = 0 & S(200) = 4. \\ S(111) = 4 & S(222) = 4 \end{array} \qquad \text{(A.31)}$$

Ad (c)

$$S(hkl) = \begin{cases} 0, & \text{if one of the } h, k, l \text{ indices is odd and two} \\ & \text{of them are even or } vice\ versa \\ 4, & \text{if all } h, k, l \text{ indices are odd or all of them} \\ & \text{are even} \end{cases} \quad .$$

$$\text{(A.32)}$$

Exercise 7.3 Four of the eight atoms within the cubic F unit cell of the diamond structure are located at $\vec{r}_1 = \vec{0}$, $\vec{r}_2 = \frac{1}{2}\vec{a}_1 + \frac{1}{2}\vec{a}_2$, $\vec{r}_3 = \frac{1}{2}\vec{a}_1 + \frac{1}{2}\vec{a}_3$, and $\vec{r}_4 = \frac{1}{2}\vec{a}_2 + \frac{1}{2}\vec{a}_3$, where $\vec{a}_1, \vec{a}_2, \vec{a}_3$ define the unit cell.

The rest of the position vectors can be expressed as the following sums

$$\vec{r}_{i+4} = \vec{r}_i + \vec{t}, \quad i = 1, \ldots, 4, \tag{A.33}$$

where $\vec{t} = \frac{1}{4}\vec{a}_1 + \frac{1}{4}\vec{a}_2 + \frac{1}{4}\vec{a}_3$. Since the first four atoms correspond to the *fcc* structure and each of the other four atoms is just shifted by the same vector $\vec{t}$ with respect to one of the *fcc* atoms, we have

$$S_{diamond} = \sum_{m=1}^{8} \exp[i 2\pi (h u_m + k v_m + l w_m)$$

$$= S_{fcc} + S_{fcc} \exp\left[i\frac{\pi}{2}(h+k+l)\right]$$

$$= S_{fcc}\left\{1 + \exp\left[i\frac{\pi}{2}(h+k+l)\right]\right\}. \tag{A.34}$$

The factor $\{1 + \exp[i\pi/2(h+k+l)]\}$ is a real number in the following cases

$$\left\{1 + \exp\left[i\frac{\pi}{2}(h+k+l)\right]\right\} = \begin{cases} 0, & if\ h+k+l = 2n,\ n\text{-}odd \\ 2, & if\ h+k+l = 2n,\ n\text{-}even \end{cases}. \tag{A.35}$$

Exercise 7.4 In order to find the expression of the atomic scattering factor f for the case when the X-ray beam is parallel to the lattice plane ($\theta = 0$), we will consider the extreme case of $\theta \to 0$. In this case, the parameter a defined in Eq. 7.36 also tends to 0 and we have

$$ar \to 0 \Rightarrow \frac{\sin(ar)}{ar} \to 1. \tag{A.36}$$

Then

$$f = \frac{4\pi}{a} \int_0^{\infty} \rho(r) r \sin(ar)\, dr$$

$$= 4\pi \int_0^{\infty} \rho(r) r^2 \frac{\sin(ar)}{ar}\, dr \xrightarrow[a \to 0]{} 4\pi \int_0^{\infty} \rho(r) r^2 dr. \tag{A.37}$$

If we now take advantage of the equality

$$4\pi = \int_0^{2\pi} d\phi \int_0^{\pi} \sin\vartheta\, d\vartheta, \tag{A.38}$$

we obtain

$$f \xrightarrow[\theta \to 0]{} \int_0^{2\pi} d\phi \int_0^\pi \sin\vartheta\, d\vartheta \int_0^\infty \rho(r)\, r^2 dr = \int \rho(r)\, d^3 r = Z. \quad \text{(A.39)}$$

That is to say, the atomic scattering factor tends toward the number Z of electrons in the atom,

$$f \xrightarrow[\theta \to 0]{} Z.$$

References

1. N. W. Ashcroft, N. D. Mermin, *Solid State Physics* (Holt, Rinehart and Winston, 1976).
2. A. Authier, *International Tables for Crystallography, Volume D: Physical Properties of Crystals* (Springer, 2003).
3. F. Cardarelli, *Materials Handbook: A Concise Desktop Reference* (Springer, 2008).
4. J. D'Ans, R. Blachnik, E. Lax, C. Synowietz, *Taschenbuch Für Chemiker und Physiker: Band 3: Elemente, Anorganische Verbindungen und Materialien, Minerale* (Springer, 1998).
5. J. R. Davis, *Metals Handbook Desk Edition* (Taylor and Francis, 1998).
6. A. S. Davydov, *Quantum Mechanics* (Pergamon Press, 1965).
7. B. Douglas, S.-M. Ho, *Structure and Chemistry of Crystalline Solids* (Springer, 2006).
8. R. T. Downs, M. Hall-Wallace, The American Mineralogist Crystal Structure Database. *American Mineralogist* **88**, 247 (2003).
9. S. Dresselhaus, G. Dresselhaus, A. Jorio, *Group Theory: Application to the Physics of Condensed Matter* (Springer, 2008).
10. P. Eckerlin, H. Kandler, *Structure Data of Elements and Intermetallic Phases* (Springer, 1971).
11. W. F. Gale, T. C. Totemeier, *Smithells Metals Reference Book* (Elsevier, 2004).
12. C. Giacovazzo, *Fundamentals of Crystallography* (Oxford University Press, 1992).
13. S. Gražulis et al., Crystallography Open Database: an open-access collection of crystal structures. *Journal of Applied Crystallography* **42**, 1 (2009).
14. T. Hahn, *International Tables for Crystallography, Volume A: Space Group Symmetry* (Springer, 2005).
15. T. Hahn, H. Wondratschek, *Symmetry of Crystals* (Heron Press, 1994).

16. M. Hellenbrandt, The Inorganic Crystal Structure Database (ICSD): present and future. *Crystallography Reviews* **10**, 17 (2004).

17. C. Kittel, *Introduction to Solid State Physics* (Wiley, 1976).

18. W.-K. Li, G.-D. Zhou, T. C. W. Mak, *Advanced Structural Inorganic Chemistry* (Oxford University Press, 2008).

19. D. R. Lide, *CRC Handbook of Chemistry and Physics* (CRC Press, 2007).

20. X. Luo et al., Predicting two-dimensional boron–carbon compounds by the global optimization method. *Journal of the American Chemical Society* **133**, 16285 (2011).

21. W. Martienssen, H. Warlimont, *Springer Handbook of Condensed Matter and Materials Data* (Springer, 2005).

22. J. W. Martin, *Concise Encyclopedia of the Structure of Materials* (Elsevier, 2007).

23. L. R. Morss, N. M. Edelstein, J. Fuger, J. J. Katz, *The Chemistry of the Actinide and Transactinide Elements* (Springer, 2006).

24. W. B. Pearson, *A Handbook of Lattice Spacings and Structures of Metals and Alloys* (Pergamon Press, 1958).

25. V. K. Pecharsky, P. Y. Zavalij, *Fundamentals of Powder Diffraction and Structural Characterization of Materials* (Springer, 2009).

26. G. S. Rohrer, *Structure and Bonding in Crystalline Materials* (Cambridge University Press, 2001).

27. D. Sands, *Introduction to Crystallography* (Courier Dover Publications, 1993).

28. U. Shmueli, *International Tables for Crystallography, Volume B: Reciprocal Space* (Springer, 2001).

29. L. Smart, E. A. Moore, *Solid State Chemistry: An Introduction* (CRC Press, 2005).

30. R. J. D. Tilley, *Crystals and Crystal Structures* (Wiley, 2006).

31. E. Y. Tonkov, E. G. Ponyatovsky, *Phase Transformations of Elements Under High Pressure* (CRC Press, 2004).

32. R. W. G. Wyckoff, *The Structure of Crystals* (Read Books, 2007).

33. D. A. Young, *Phase Diagrams of the Elements* (University of California Press, 1991).

Binary Compounds Index

Index